SUMMER LINK MATH

SUMMER BEFORE GRADE 1

McGraw Hill Children's Publishing
Columbus, Ohio

Summer Link

McGraw Hill Children's Publishing

Copyright © 2004 McGraw-Hill Children's Publishing. Published by American Education Publishing, an imprint of McGraw-Hill Children's Publishing, a Division of The McGraw-Hill Companies.

Printed in the United States of America. All rights reserved. Except as permitted under the United States Copyright Act, no part of this publication may be reproduced or distributed in any form or by any means, or stored in a database or retrieval system, without prior written permission from the publisher, unless otherwise indicated.

Send all inquiries to:
McGraw-Hill Children's Publishing
8787 Orion Place
Columbus, OH 43240-4027

ISBN 0-7696-3310-2

1 2 3 4 5 6 7 8 9 10 QPD 09 08 07 06 05 04

The McGraw-Hill Companies

Table of Contents

Counting 1–20 4–30
Number Recognition 31
Number Words 32–33
Sequencing Numbers 34–35
Ordinal Numbers 36–38
Equal, More, and Less 39–43
Addition 44–47
Subtraction 48–49
Addition and Subtraction 50–53
Time 54–56
Money 57–64
Shapes 65–76
Patterns 77–78
Graphing 79–84
Answer Key 85–93
Developmental Skills for First Grade Success 94

Number Recognition

Directions: Write the numbers 1–10. Color the bear.

1 2 3 4 5 6 7 8 9 10

Zero

Directions: Color the tank to show that it has 0 fish. Color the tanks that have 0 fish.

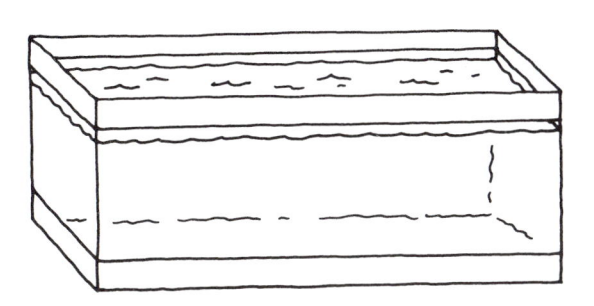

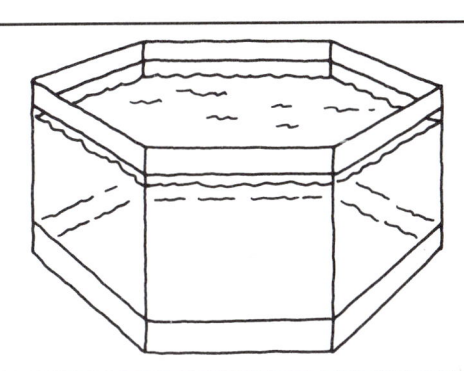

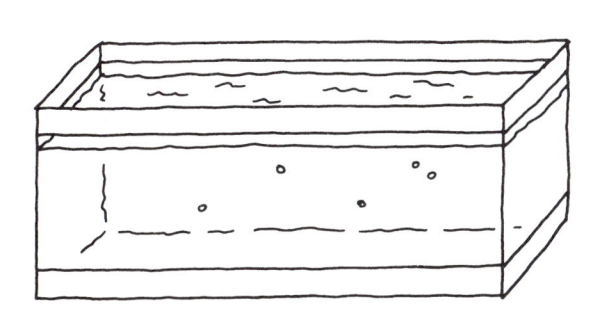

One and Two

Directions: Count how many cars are on each track. Circle the number that shows how many.

Three

Directions: Color the 3 kittens in the basket. Color 3 animals in each group.

Four

Directions: Color the 4 crayons. Count how many. Circle the correct number.

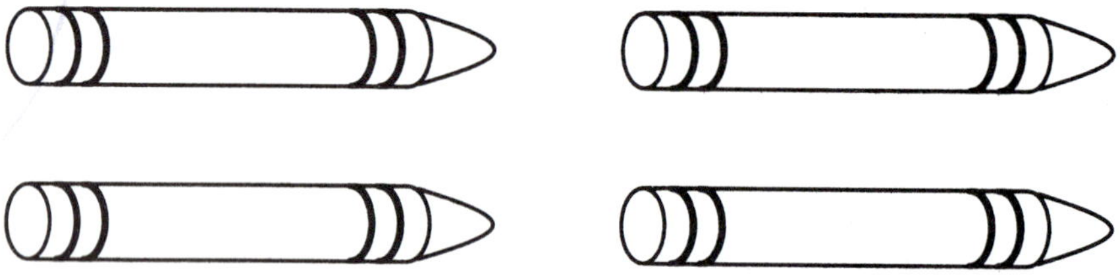

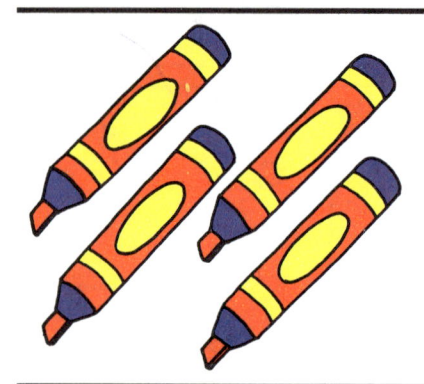

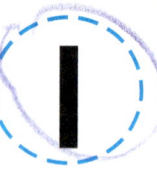

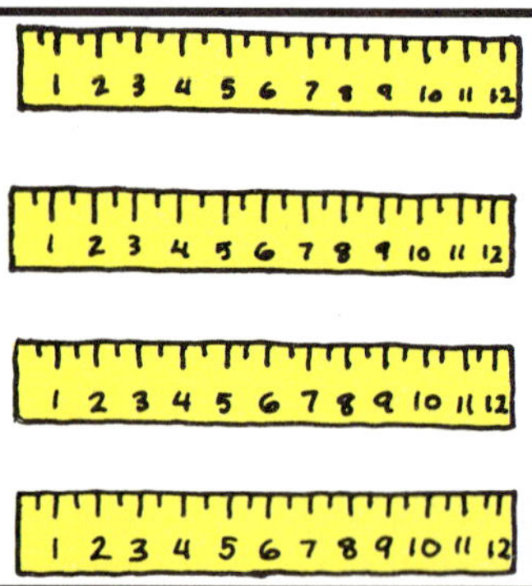

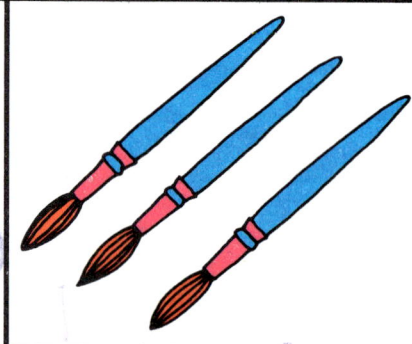

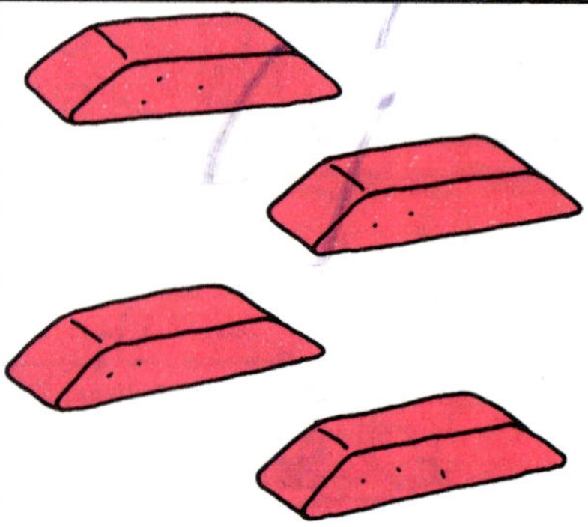

Five

Directions: Color the 5 party hats. Color and circle the groups that have 5.

Review Numbers 1–5

Directions: Look at the picture. Read the questions. Circle the correct number.

◆ How many 🦋 in all?   3   4   5

◆ How many 🐝 in all?   3   4   5

◆ How many 🐞 in all?   3   4   5

Summer Link Math Grade 1

Review Numbers 1-5

Directions: Draw a line from the number to the group that matches it.

1

2

3

4

5

Summer Link Math Grade 1

Six

Directions: Look at the number 6. Count the teddy bears. Trace the circle to show this is a group of 6. Circle the group if it shows 6.

Review Numbers 1-6

Directions: Count each group of blocks. Trace each number. Count each group of blocks below. Write the number.

1   2   3   4   5   6

___ ___ ___ ___ ___ ___

Seven

Directions: Look at the number 7. Count how many bees. Color the 7 bees. Count how many. Circle the number.

Eight

Directions: Look at the number 8. Count the envelopes. Trace the circle to show this is a group of 8. Circle the group if it shows 8.

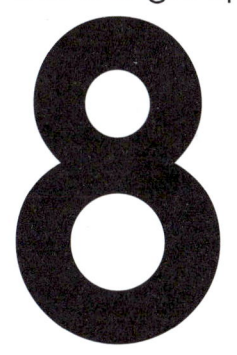

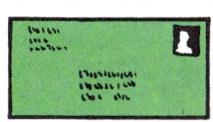

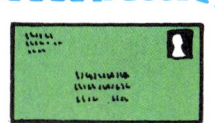

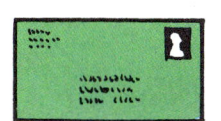

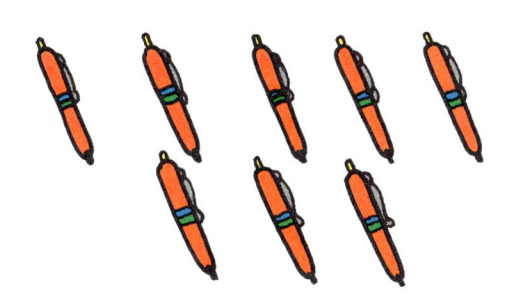

Nine

Directions: Look at the number 9. Circle 9 cars. Circle the signs to show the number.

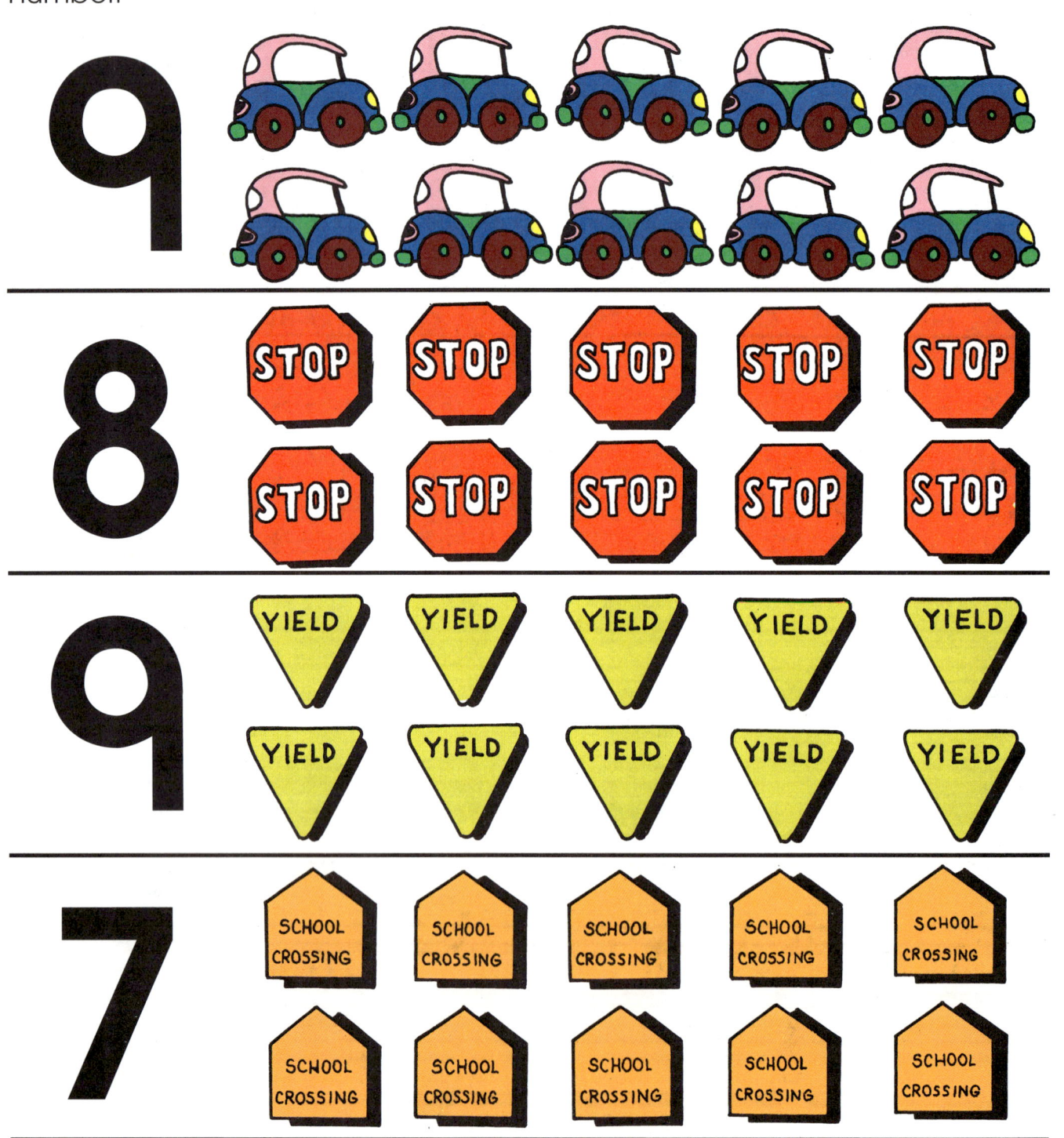

9

8

9

7

Ten

Directions: Look at the number 10. Trace the circle to show this is a group of 10. Circle each group of 10 objects.

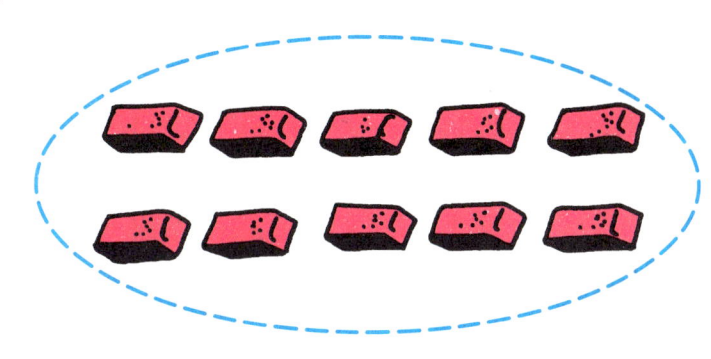

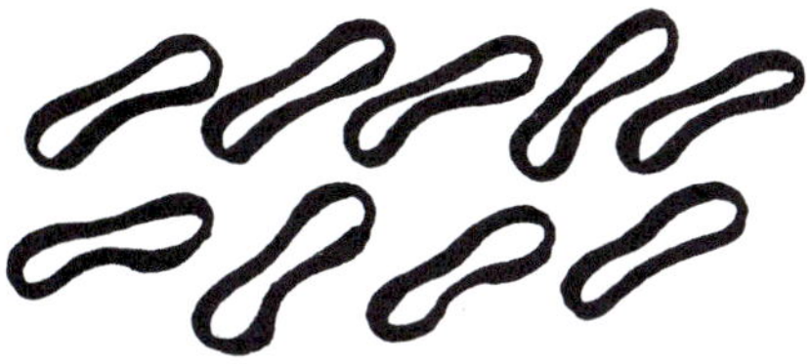

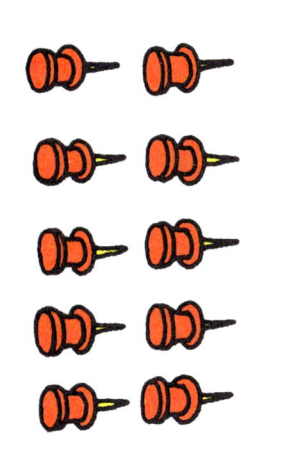

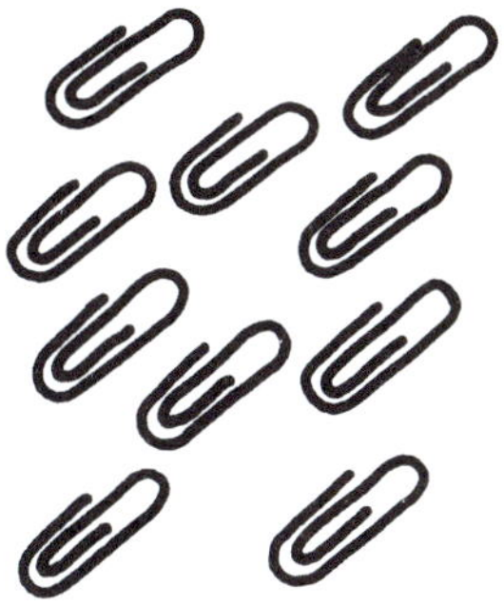

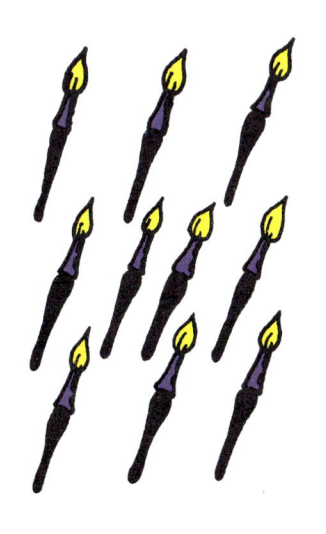

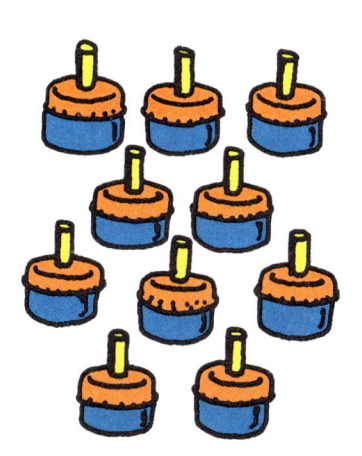

Nine and Ten

Directions: Count and write the number in each box. Circle the groups of 9. Color the groups of 10.

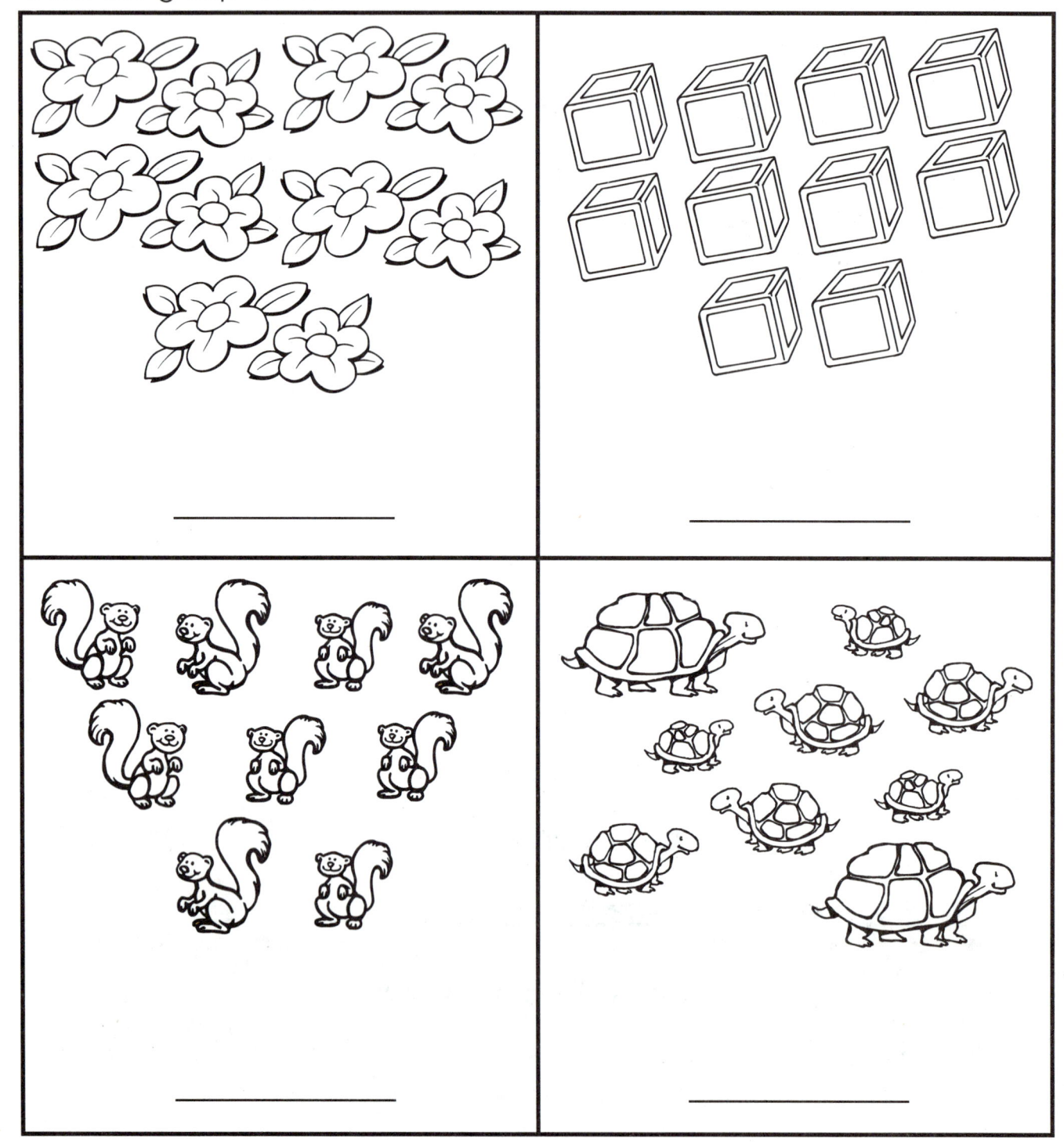

Name _____

Review Numbers 7-10

Directions: Count each group of balloons. Trace each number. Count each group of balloons below. Write the number.

7   8   9   10

____   ____   ____   ____

19   Summer Link Math Grade 1

Eleven and Twelve

Directions: Trace and write the numbers 11 and 12. Count and write the numbers.

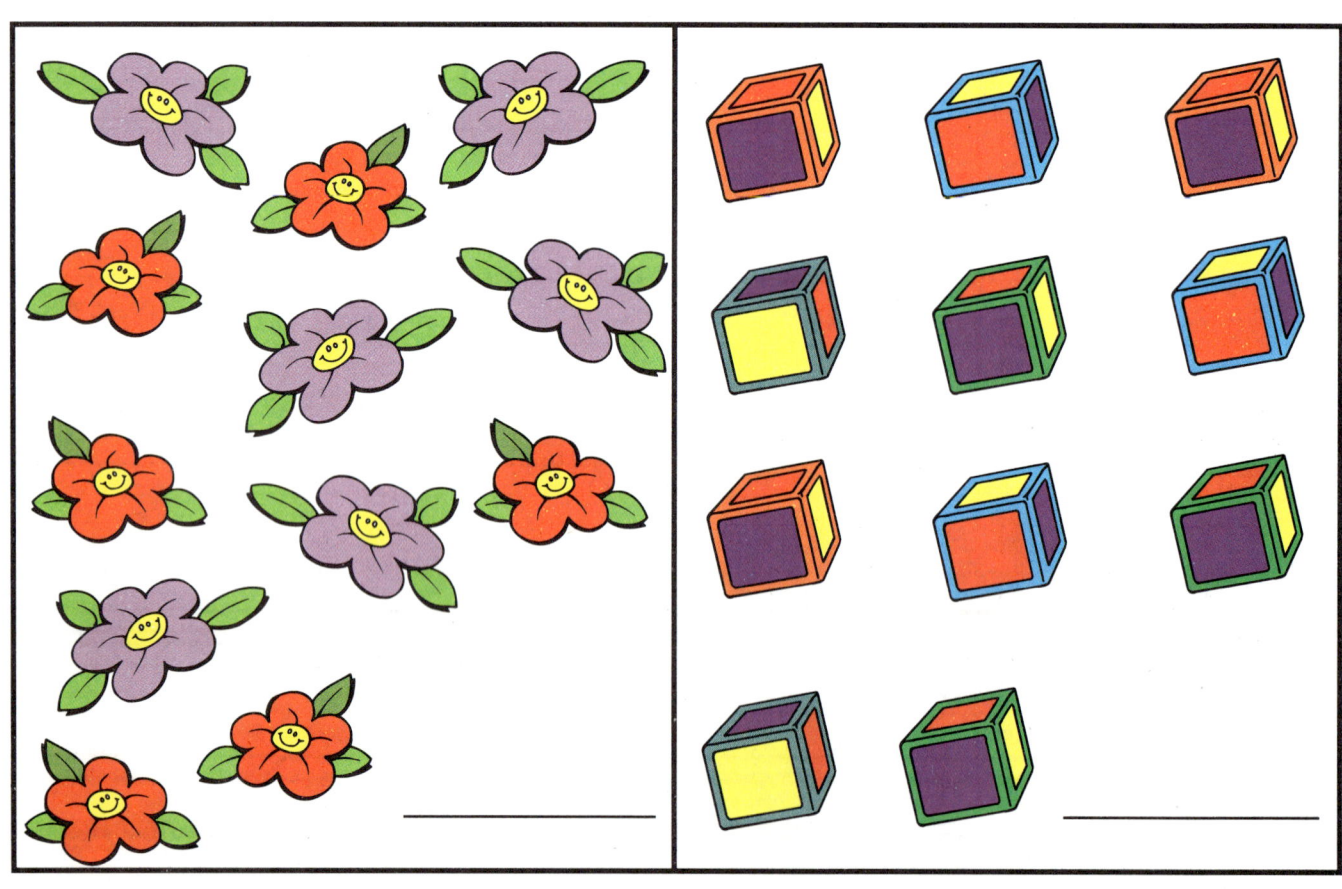

Summer Link Math Grade 1

Eleven and Twelve

Directions: Draw flowers to show the number in each box.

11

12

Review Numbers 1-12

Directions: Count the number of colored squares. Then write the correct number.

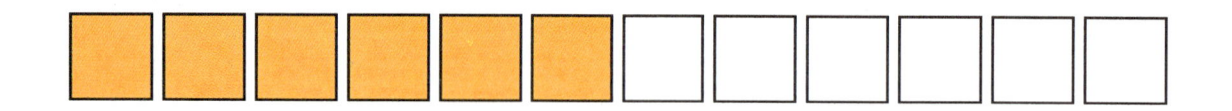

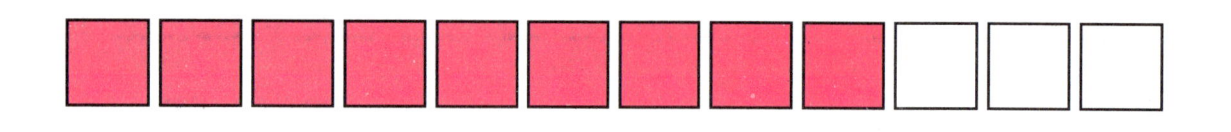

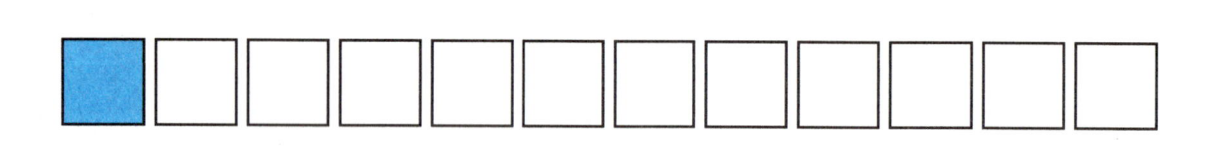

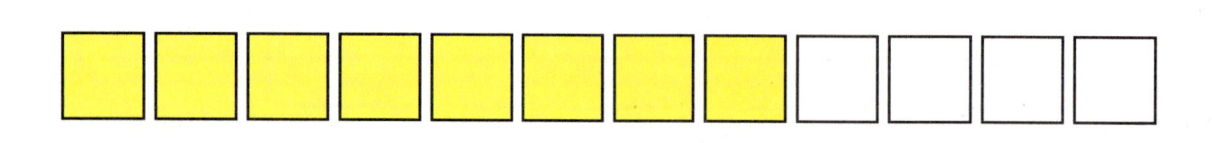

_____

_____

_____

_____

_____

Name _____

Thirteen

Directions: Trace and write the number 13. Complete each puzzle by writing or drawing the missing number of flowers.

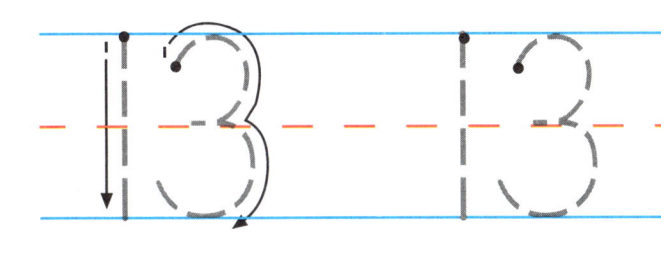

13

Name _____

Fourteen

Directions: Trace and write the number 14. Connect the dots. Color the picture. What is it?

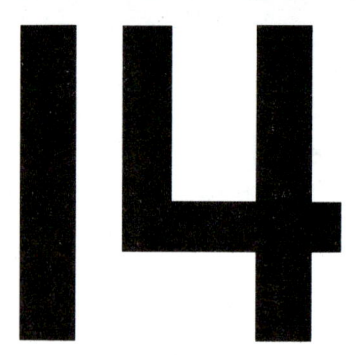

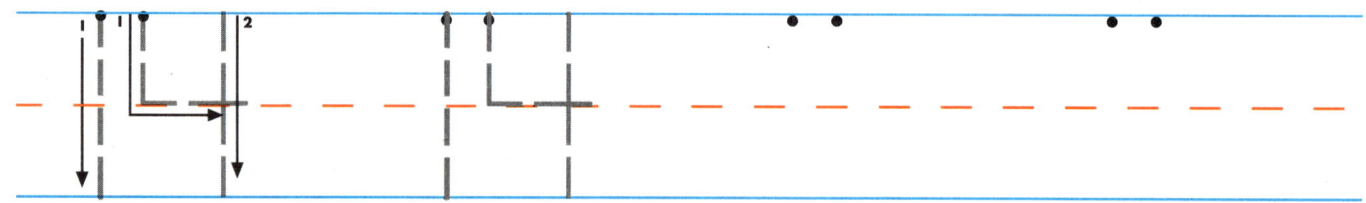

14

Summer Link Math Grade 1   24

Fifteen

15

Directions: Trace and write the number 15. Write the missing pool ball numbers.

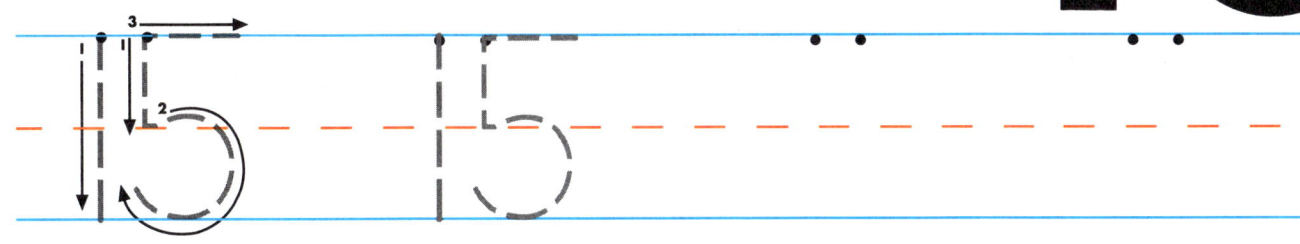

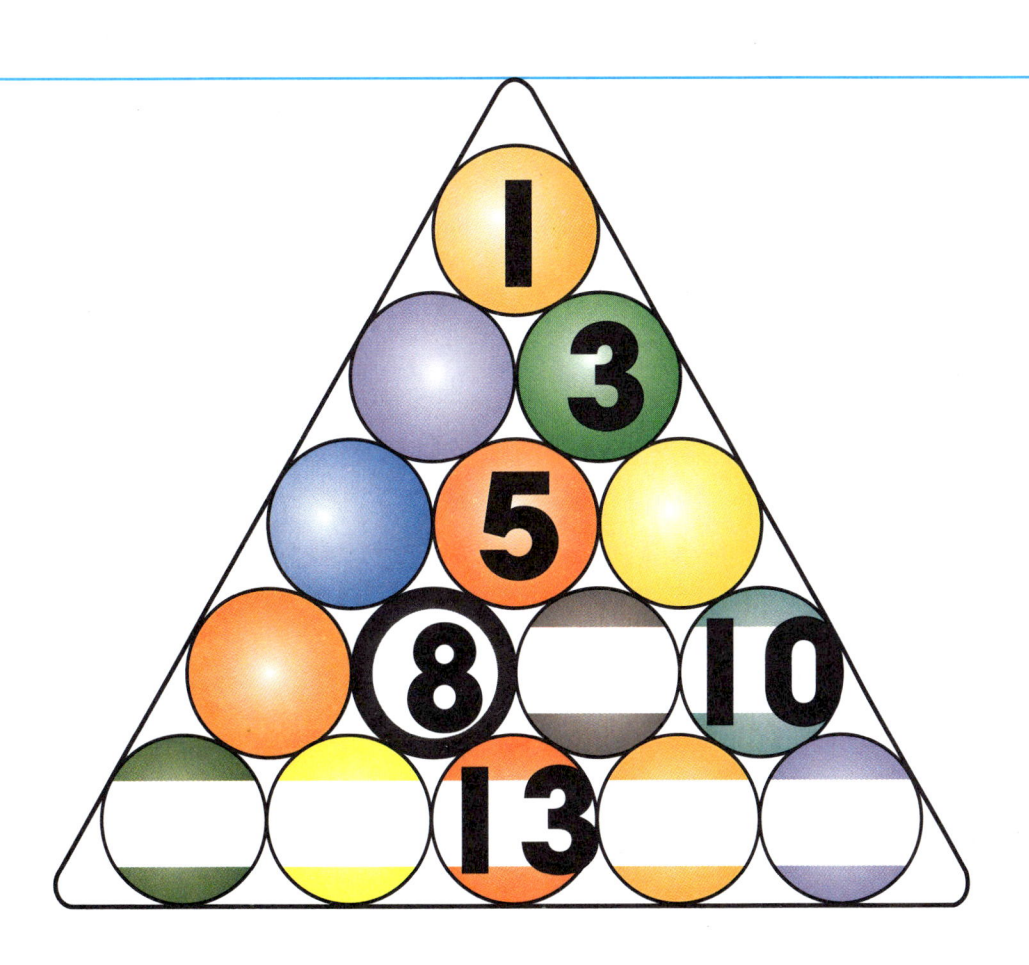

Name _____

Sixteen

16

Directions: Trace and write the number 16. Draw eight legs on each spider.

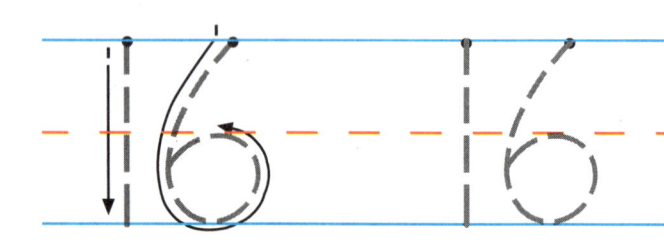

How many legs are there in all? _____

Summer Link Math Grade 1

Name _____

Seventeen

17

Directions: Trace and write the number 17. Circle each group of 17 things. Color the dog.

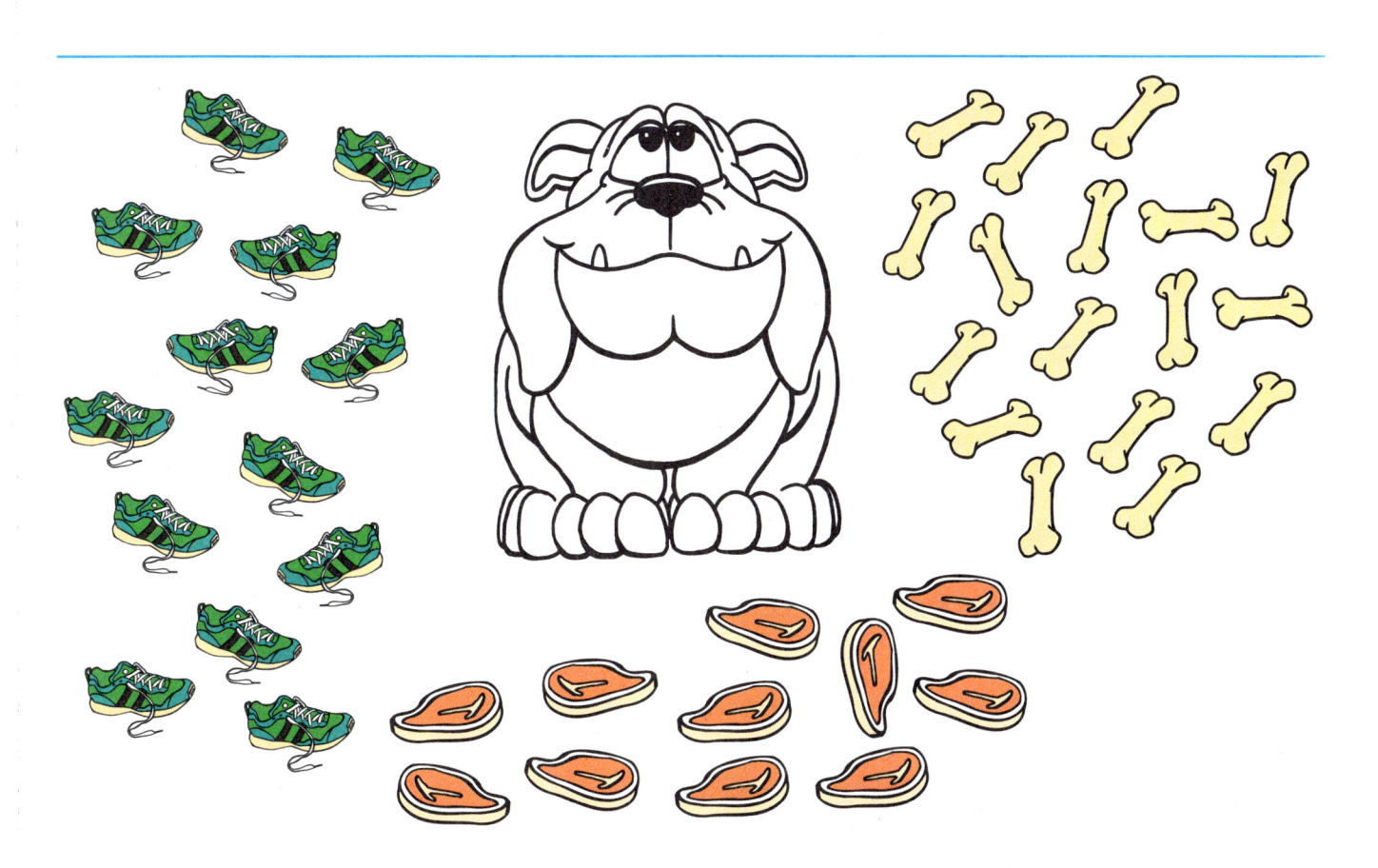

27 Summer Link Math Grade 1

Eighteen

Directions: Trace and write the number 18. Help Filbert Fish find his way to the top. Write the numbers 1–18 in each bubble along the way.

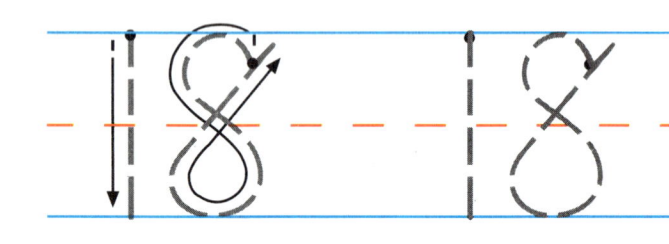

18

Nineteen

19

Directions: Trace and write the number 19. Circle the numbers 1–19 in the picture.

Name _____

Twenty

20

Directions: Trace and write the number 20. Connect the dots to find the hidden picture. What is it?

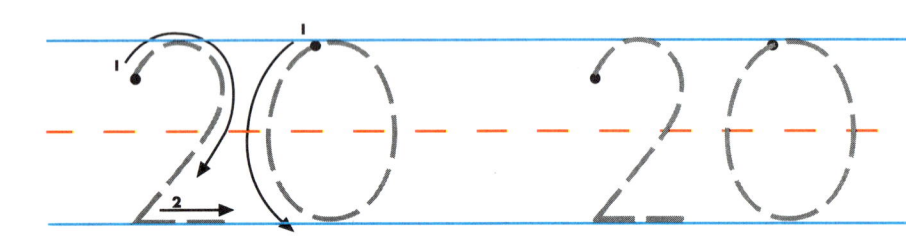

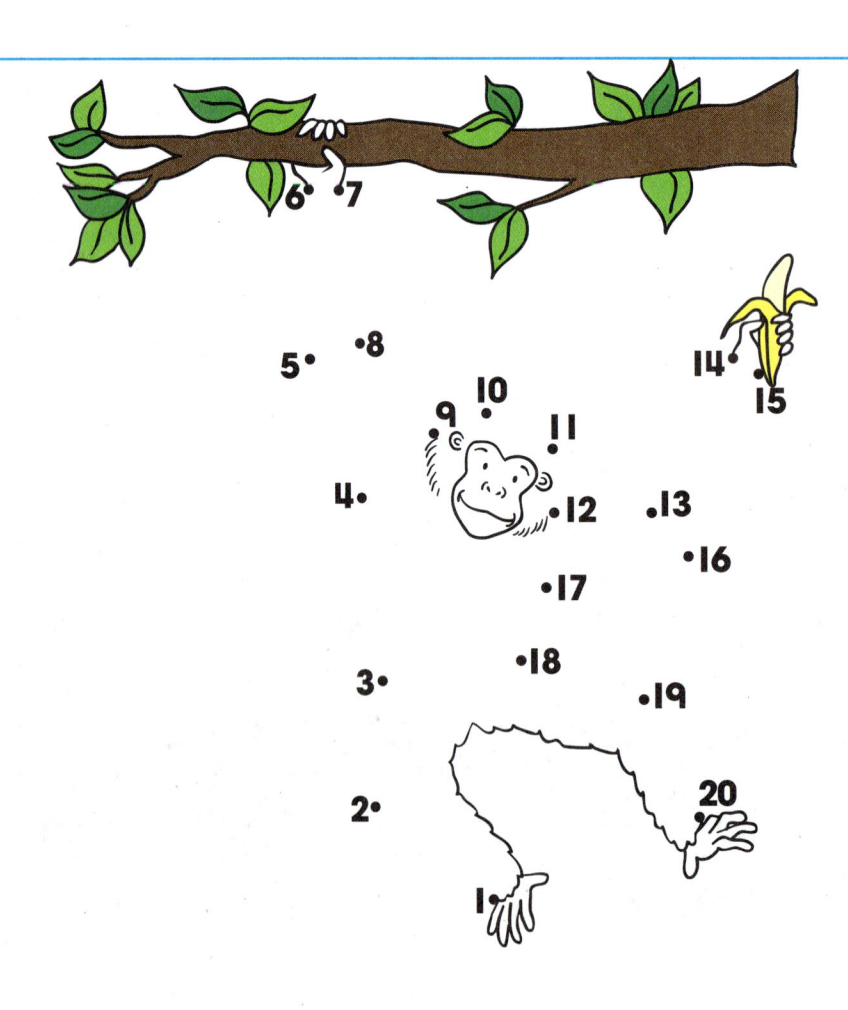

Summer Link Math Grade 1

Number Recognition

Directions: Count the number of objects in each group. Draw a line to the correct number.

1
2
3
4
5
6
7
8
9
10

Number Words

Directions: Number the buildings from one to six.

Directions: Draw a line from the word to the number.

two   1

five   3

six   5

four   6

one   2

three   4

Summer Link Math Grade 1

Number Words

Directions: Number the buildings from five to ten.

Directions: Draw a line from the word to the number.

nine   **8**

seven   **10**

five   **7**

eight   **5**

six   **9**

ten   **6**

Name _____

Sequencing Numbers

Sequencing is putting numbers in the correct order.

1, 2, 3, 4, 5, 6, 7, 8, 9, 10

Directions: Write the missing numbers.

Example: 4, __5__, 6

3, _____ ,5   7, _____ ,9   8, _____ ,10

6, _____ ,8   _____ ,3, 4   _____ ,5, 6

5, 6, _____   _____ ,6, 7   _____ ,3, 4

_____ ,4, 5   _____ ,7, 8   5, _____ ,7

2, 3, _____   1, 2, _____   7, 8, _____

2, _____ ,4   _____ ,2, 3   4, _____ ,6

6, 7, _____   3, 4, _____   1, _____ ,3

7, 8, _____   _____ ,3, 4   _____ ,9, 10

Summer Link Math Grade 1

34

Review

Directions: Count the objects and write the number.

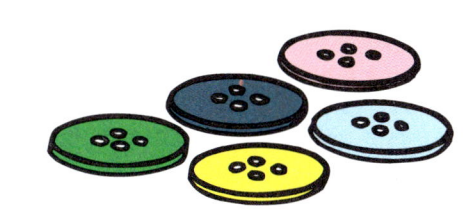

----------   ----------   ----------

Directions: Match the number to the word.

two   1

four   9

seven   2

three   3

one   4

nine   7

Ordinal Numbers

Ordinal numbers are used to indicate order in a series, such as **first**, **second**, or **third**.

Directions: Draw a line to the picture that goes with the ordinal number in the left column.

eighth

third

sixth

ninth

seventh

second

fourth

first

fifth

tenth

Ordinal Numbers

Directions: Color the first leaf red. Circle the third leaf.

Directions: Color the fourth balloon purple. Draw a line under the second balloon.

Name _____

Orderly Ordinals

Directions: Write each word on the correct line to put the words in order.

| second | fifth | seventh | first | tenth |
| third | eighth | sixth | fourth | ninth |

1. _____   6. _____
2. _____   7. _____
3. _____   8. _____
4. _____   9. _____
5. _____   10. _____

Directions: Which picture is circled in each row? Underline the word that tells the correct number.

third   fourth

fourth   sixth

first   ninth

third   fifth

fifth   sixth

second   third

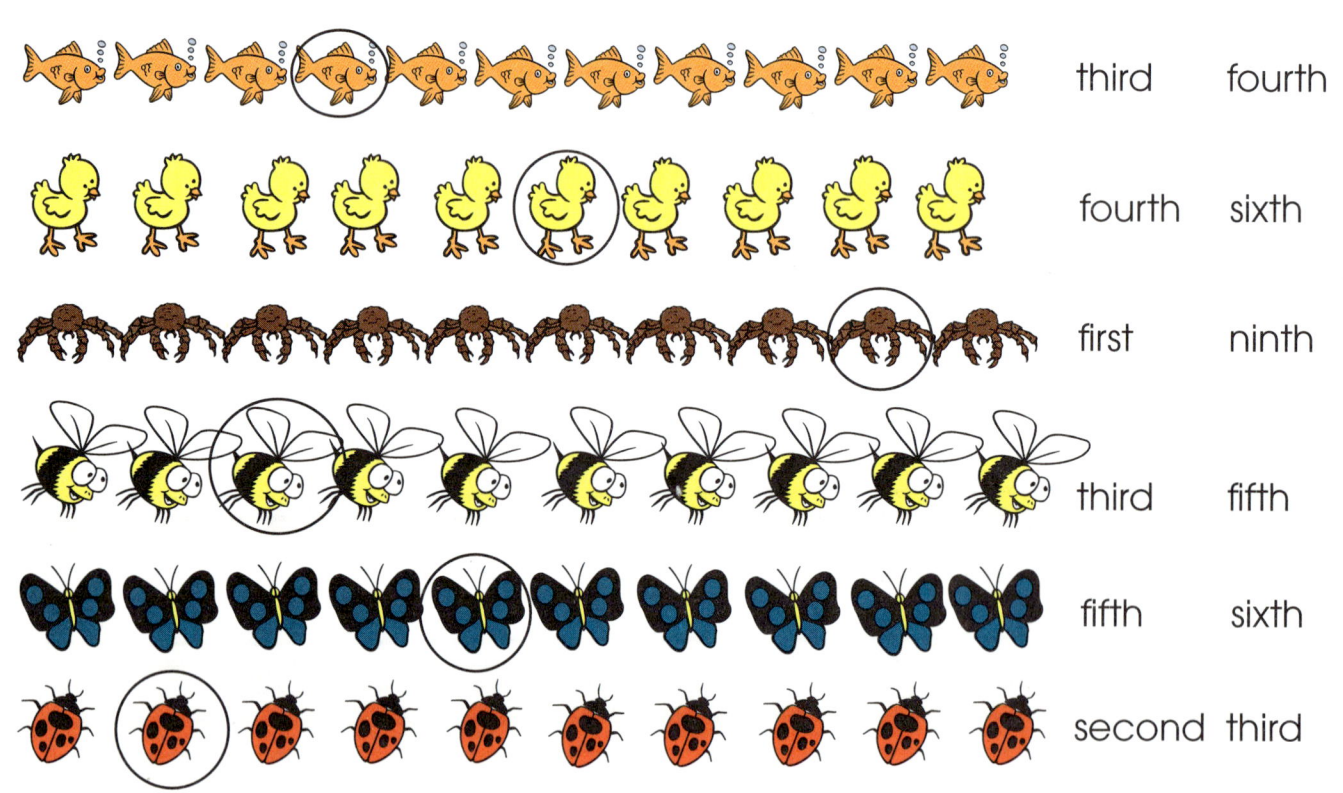

Summer Link Math Grade 1

One for Each

Directions: Each circus seal needs one ball. Draw a ball for each seal.

More

Directions: Circle the group in each box that has more.

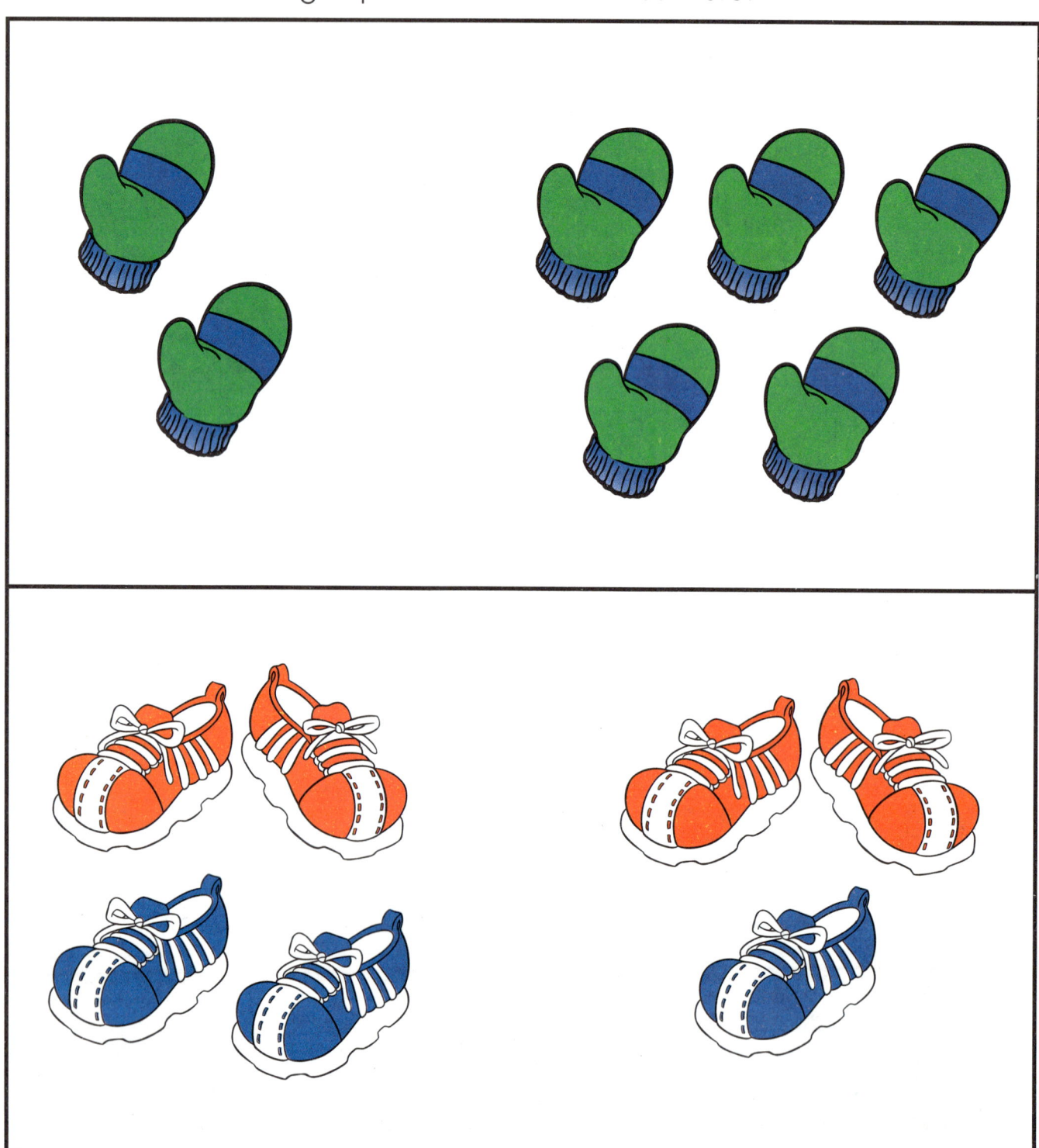

More

Directions: Trace the lines to match the rows of mittens one-to-one. Circle the 6 to show that 6 is more than 4. Match the objects one-to-one. Circle the number that is more.

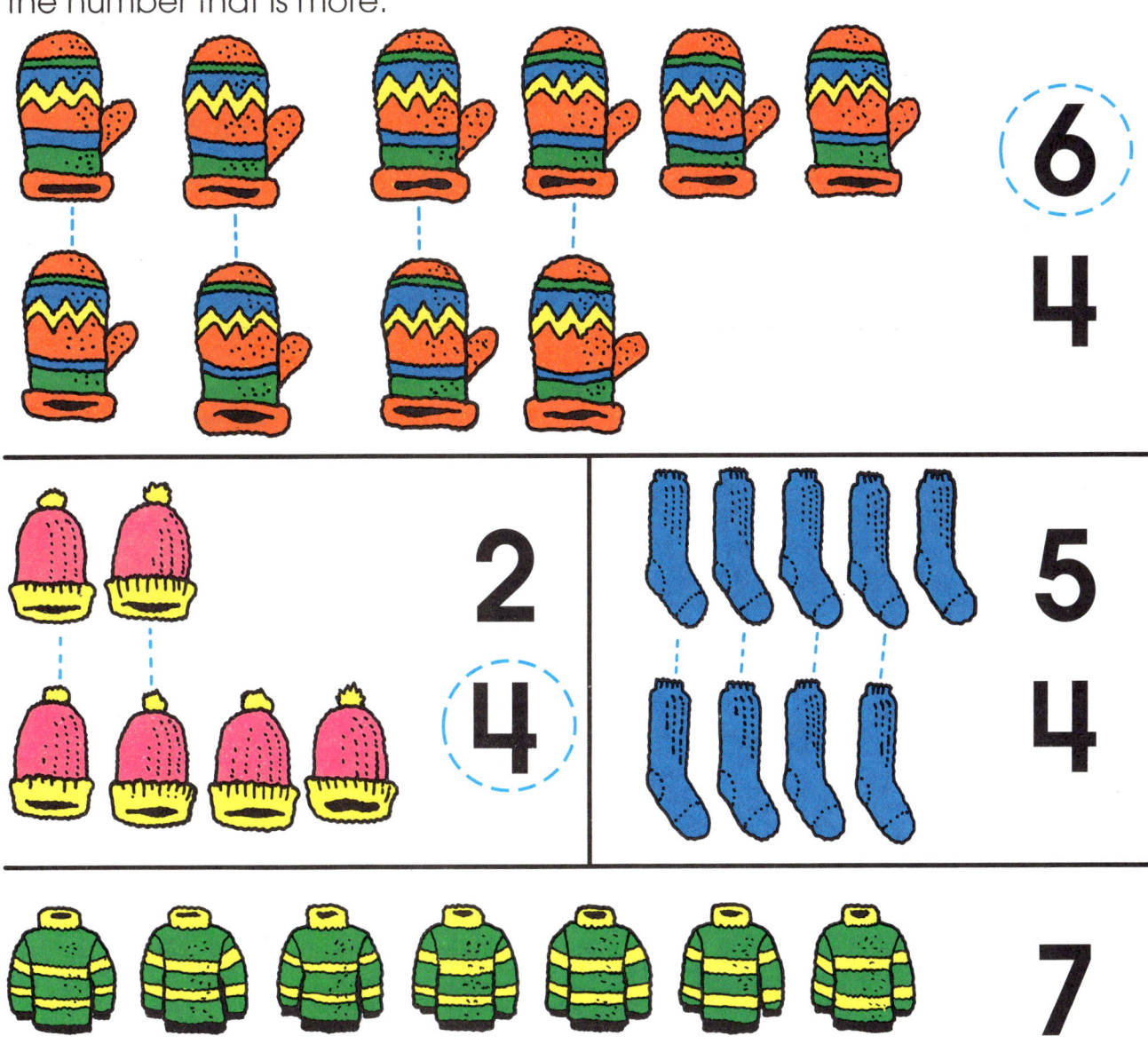

Name _____

Less

Directions: Circle the group in each box that has less.

Less

Directions: Trace the lines to match the fish one-to-one. Circle the 4 to show that 4 is less than 6. Match the objects one-to-one. Circle the number that is less.

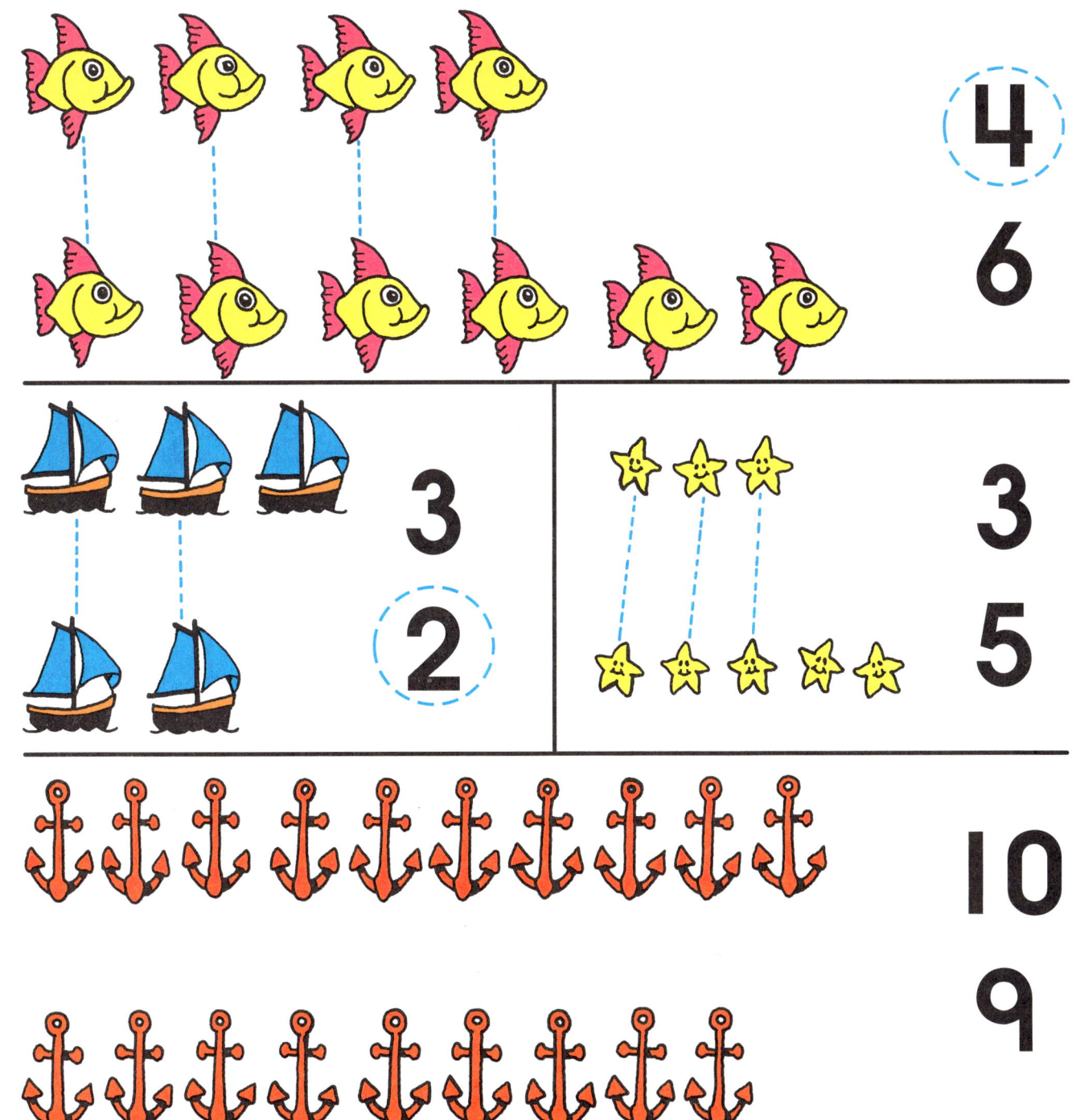

Addition 1, 2

Addition means "putting together" or adding two or more numbers to find the sum. "+" is a plus sign. It means to add the 2 numbers. "=" is an equals sign. It tells how much they are together.

Directions: Count the cats and tell how many.

Summer Link Math Grade 1

Name _____

Addition

Directions: Count the shapes and write the numbers below to tell how many in all.

Name _____

Picture Problems: Addition

Directions: Solve the number problem under each picture.

6 + 2 = ___

3 + 1 = ___

5 + 3 = ___

1 + 7 = ___

4 + 5 = ___

0 + 7 = ___

Summer Link Math Grade 1

46

How Many in All?

Directions: Count the number in each group and write the number on the line. Then, add the groups together and write the sum.

_____ strawberries

_____ strawberries

How many in all? _____

_____ cookies

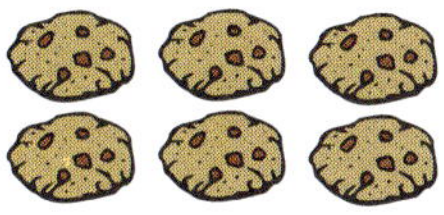

_____ cookies

How many in all? _____

_____ shoes

_____ shoes

How many in all? _____

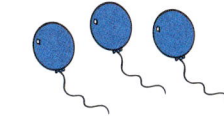

_____ balloons

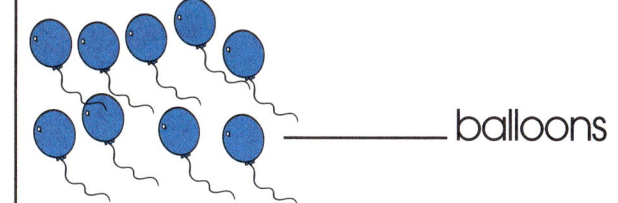

_____ balloons

How many in all? _____

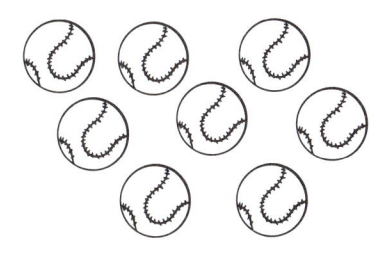

_____ balls

_____ balls

How many in all? _____

_____ flowers

_____ flowers

How many in all? _____

Name _____

Subtraction 1, 2, 3

Subtraction means "taking away" or subtracting one number from another. "−" is a minus sign. It means to subtract the second number from the first.

Directions: Practice writing the numbers and then subtract. Draw dots and cross them out, if needed.

1

2

3

$$\begin{array}{r} 3 \\ -1 \\ \hline 2 \end{array}$$

$$\begin{array}{r} 4 \\ -3 \\ \hline \end{array}$$

$$\begin{array}{r} 2 \\ -1 \\ \hline \end{array}$$

$$\begin{array}{r} 3 \\ -2 \\ \hline \end{array}$$

Summer Link Math Grade 1

48

Picture Problems: Subtraction

Directions: Solve the number problem under each picture.

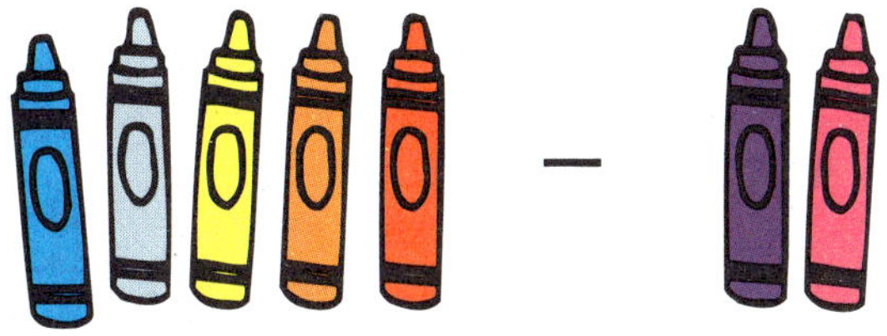

5 - 2 = ___

6 - 1 = ___

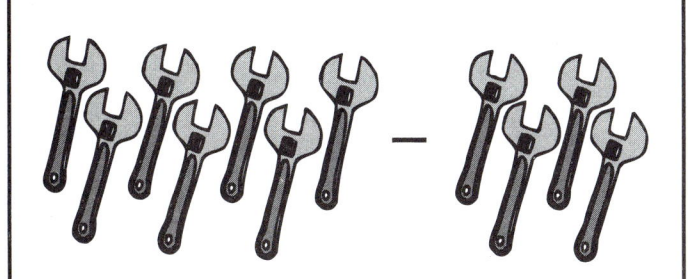

7 - 4 = ___

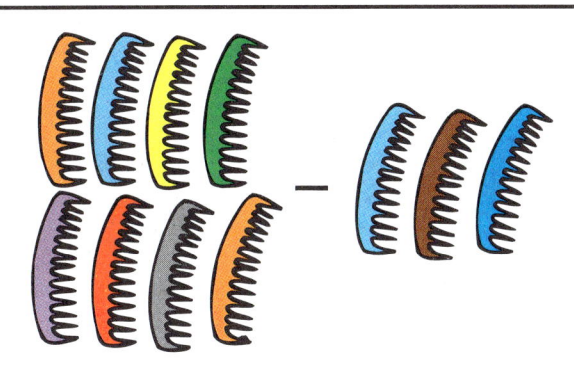

8 - 3 = ___

9 - 2 = ___

4 - 4 = ___

Picture Problems: Addition and Subtraction

Directions: Solve the number problem under each picture.

7 - 4 = ___

1 + 4 = ___

3 + 5 = ___

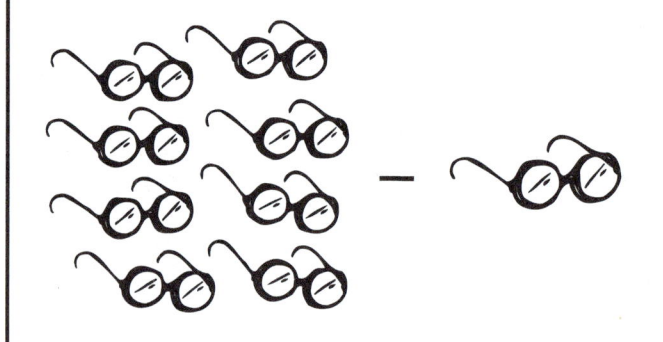

8 - 1 = ___

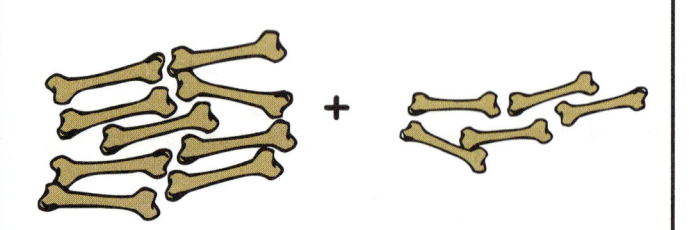

9 + 5 = ___

6 - 3 = ___

Summer Link Math Grade 1

Review: Addition and Subtraction

Directions: Solve the number problem under each picture. Write + or – to show if you should add or subtract.

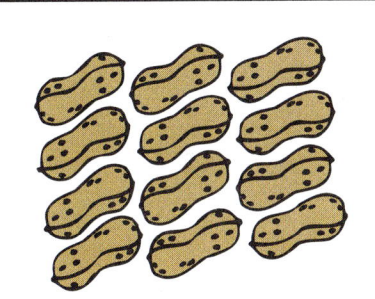

How many 🥜s are left?
12   4 = _____

How many 🍊s in all?
6   8 = _____

How many 🐱s are left?
4   4 = _____

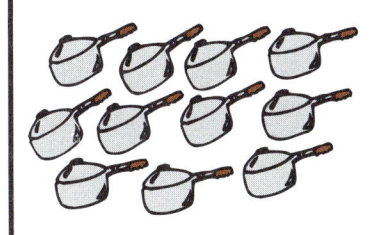

How many 🍳s are left?
11   7 = _____

How many 🎈s in all?
9   3 = _____

How many 🤡s in all?
10   0 = _____

51   Summer Link Math Grade 1

Name _____

Addition and Subtraction

Directions: Solve the problems. Remember, addition means "putting together" or adding two or more numbers to find the sum. Subtraction means "taking away" or subtracting one number from another.

1 + 3 = ___   4 - 3 = ___   4 + 5 = ___

6 + 1 = ___   7 - 2 = ___   8 - 4 = ___

                 9 - 1 = ___   10 - 3 = ___

                 5 - 2 = ___   6 + 3 = ___

                 8 + 2 = ___   5 + 5 = ___

Summer Link Math Grade 1

Color Fruit

Directions: Solve the addition and subtraction sentences below. Use the code to color the fruit.

3 — yellow	5 — orange	7 — yellow	9 — red
4 — red	6 — purple	8 — green	10 — brown

9
-4

3
$+7$

6
-3

1
$+3$

9
-2

7
$+2$

9
-1

6
$+3$

8
-2

Name _____

Time

Directions: Trace the numbers 1–12 in order on the clock.

Hickory Dickory Dock,
The mouse ran up the clock.
The clock struck one and down he ran.
Hickory Dickory Dock.

Time

Directions: Write the time that is on each clock.

Example:

___2___ o'clock

_____ o'clock

_____ o'clock

_____ o'clock

Name _____

Time

Directions: Write the time that is on each clock.

_____ o'clock

_____ o'clock

_____ o'clock

_____ o'clock

Summer Link Math Grade 1

Name _____

Pennies

Directions: A penny is worth 1¢. It is brown. Circle the correct amount of money in each row below.

Example:

1¢   (2¢)   3¢

1¢   2¢   3¢

5¢   6¢   7¢

7¢   8¢   9¢

57

Pennies

Directions: Circle the correct amount of money in each row below.

2¢   3¢   4¢

1¢   2¢   3¢

4¢   5¢   6¢

7¢   8¢   9¢

Nickels

Directions: A nickel is worth 5¢. It is silver. Circle the correct amount of money in each row below.

Example:

5¢ = 5¢

4¢   5¢   6¢

1¢   2¢   3¢

1¢   2¢   3¢

Nickels

Directions: Circle the correct amount of money in each row below.

🪙🪙🪙🪙	3¢ 4¢ 5¢	
🪙 (nickel)	4¢ 5¢ 6¢	
🪙🪙🪙	3¢ 4¢ 5¢	
🪙 (nickel) 🪙 (penny)	4¢ 5¢ 6¢	

Summer Link Math Grade 1

Dimes

Directions: A dime is worth 10¢. It is silver. Circle the correct amount of money in each row below.

Example:

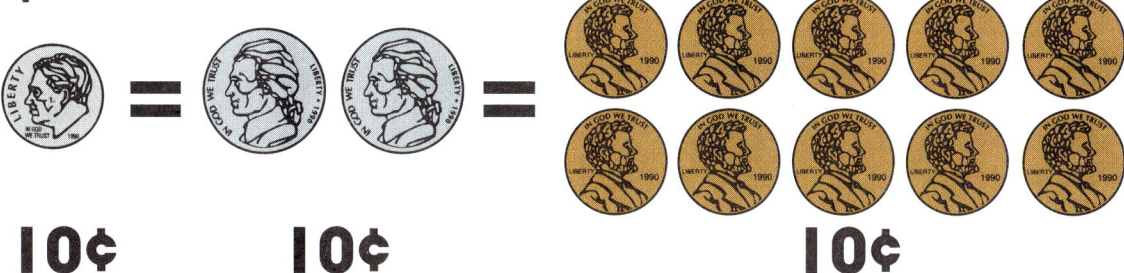

⬤ = ⬤ ⬤ = ⬤⬤⬤⬤⬤⬤⬤⬤⬤⬤

10¢   10¢   10¢

1¢   5¢   10¢

5¢   7¢   10¢

8¢   9¢   10¢

Dimes

Directions: Circle the correct amount of money in each row below.

2¢   3¢   4¢

5¢   6¢   7¢

8¢   9¢   10¢

9¢   10¢   11¢

Review Money

Directions: Match the price of each thing to the correct amount of money.

Review Money

Directions: Match the coins to the correct amount of money.

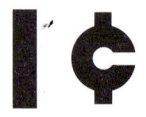

10¢

5¢

2¢

6¢

1¢

8¢

Shapes: Square

A square is a figure with four corners and four sides of the same length.

This is a square ☐ .

Directions: Find the squares and circle them.

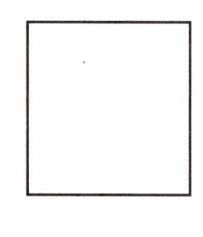

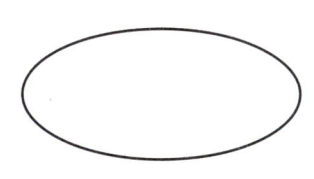

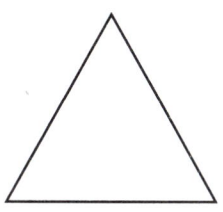

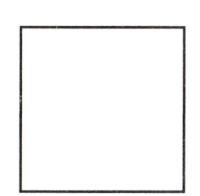

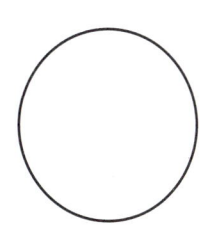

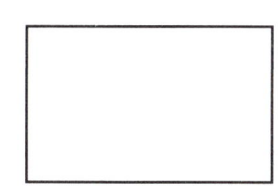

Directions: Trace the word. Write the word.

square

Shapes: Circle

A circle is a figure that is round. This is a circle ○.

Directions: Find the circles and put a square around them.

Directions: Trace the word. Write the word.

circle

Shapes: Square and Circle

Directions: Practice drawing squares. Trace the samples and make four of your own.

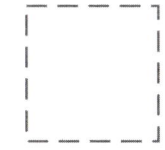

Directions: Practice drawing circles. Trace the samples and make four of your own.

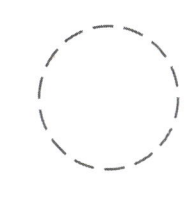

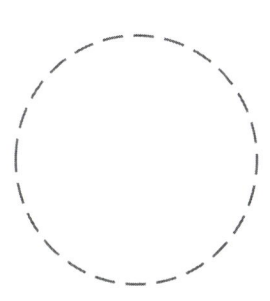

Shapes: Triangle

A triangle is a figure with three corners and three sides.

This is a triangle △ .

Directions: Find the triangles and put a circle around them.

Directions: Trace the word. Write the word.

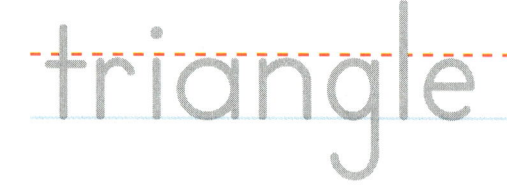

triangle

Summer Link Math Grade 1

Name _____

Shapes: Rectangle

A rectangle is a figure with four corners and four sides. Sides opposite each other are the same length.

This is a rectangle ▭.

Directions: Find the rectangles and put a circle around them.

Directions: Trace the word. Write the word.

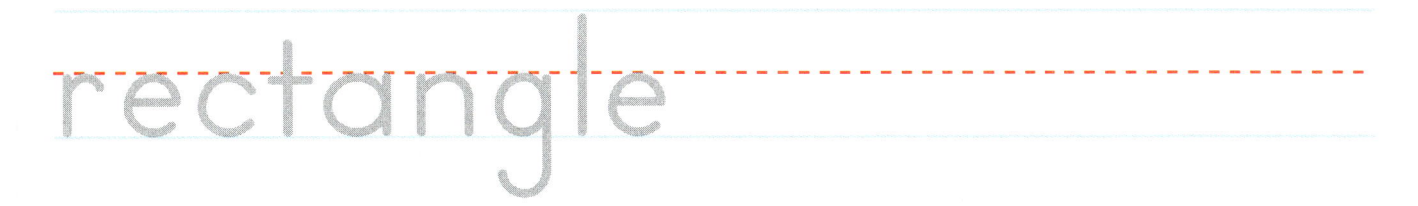

rectangle

69 Summer Link Math Grade 1

Shapes: Triangle and Rectangle

Directions: Practice drawing triangles. Trace the samples and make four of your own.

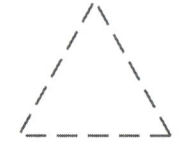

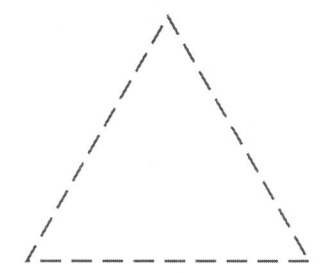

Directions: Practice drawing rectangles. Trace the samples and make four of your own.

Summer Link Math Grade 1

Shapes: Oval and Diamond

An oval is an egg-shaped figure. A diamond is a figure with four sides of the same length. Its corners form points at the top, sides, and bottom.

This is an oval ⬭.   This is a diamond ◇.

Directions: Color the ovals red. Color the diamonds blue.

Directions: Trace the words. Write the words.

oval

diamond

Name _____

Shapes: Oval and Diamond

Directions: Practice drawing ovals. Trace the samples and make four of your own.

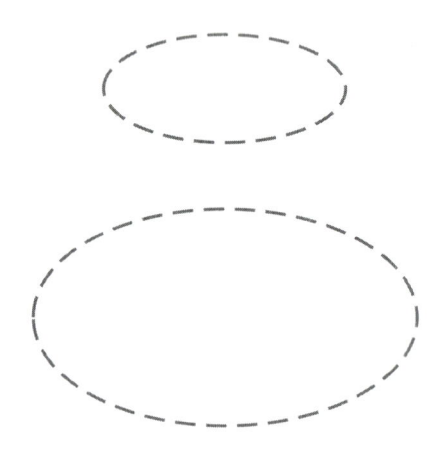

Directions: Practice drawing diamonds. Trace the samples and make four of your own.

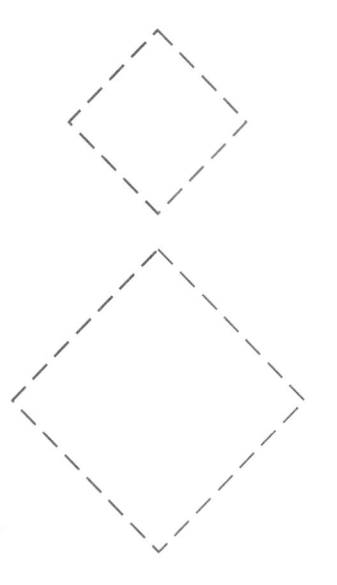

Summer Link Math Grade 1

Animal Shapes

Directions: Color: squares — green rectangles — yellow
circles — red triangles — blue

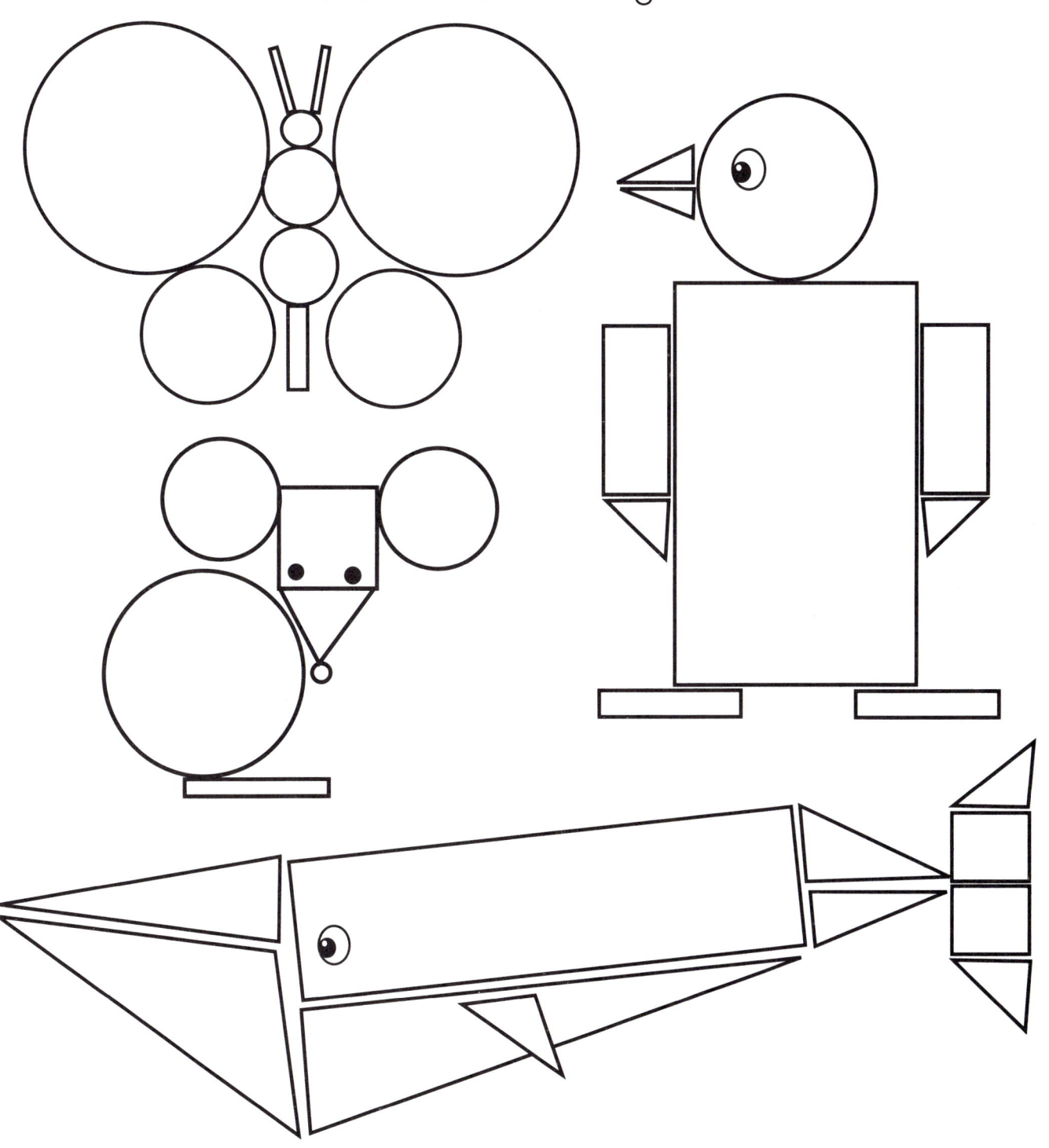

Name _____

Classifying: Shapes

Mary and Rudy are taking a trip into space. Help them find the stars, moons, circles, and diamonds.

Directions: Color the shapes.

Use yellow for ☆s.   Use blue for ☾s.

Use red for ○s.   Use purple for ◇s.

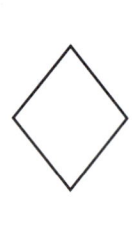

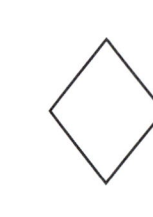

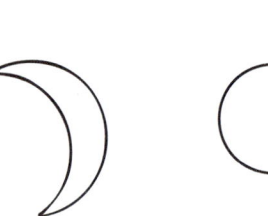

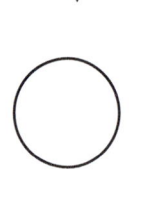

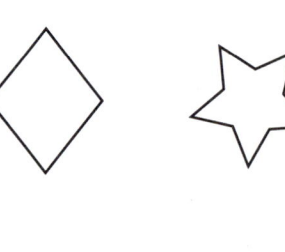

How many stars? _____   How many moons? _____

How many circles? _____   How many diamonds? _____

Summer Link Math Grade 1

Same Shape

Directions: Look at the round shapes. They are all the same shape. Draw a line from each shape in the bottom row to the box with the same shape.

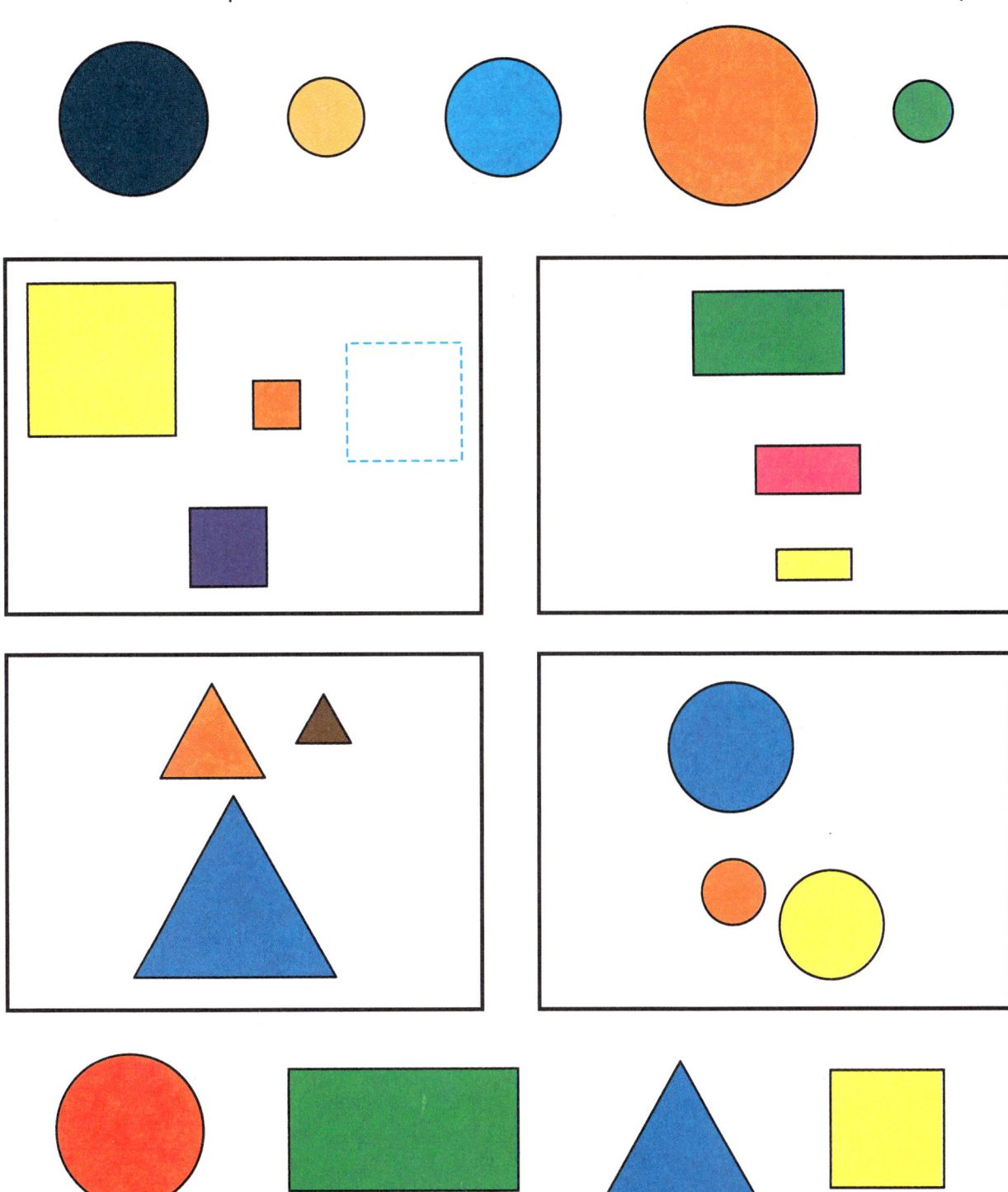

Different

Directions: Circle the shape that is different. Circle the object in each row that is different.

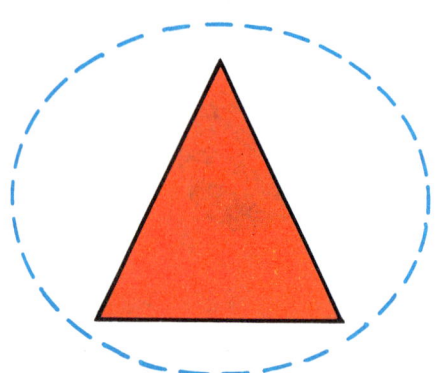

Patterns

Directions: Draw what comes next in each pattern.

Example:

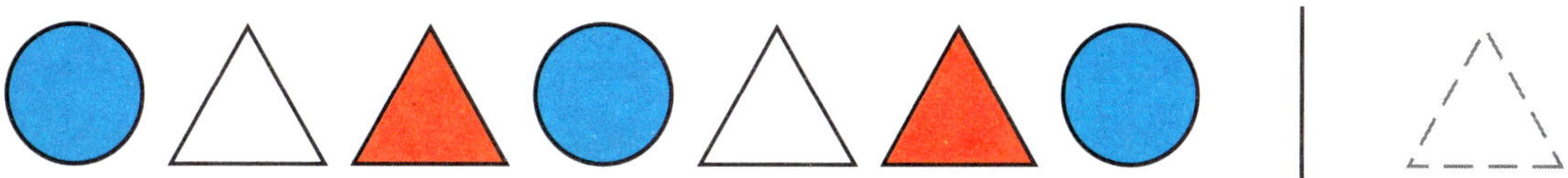

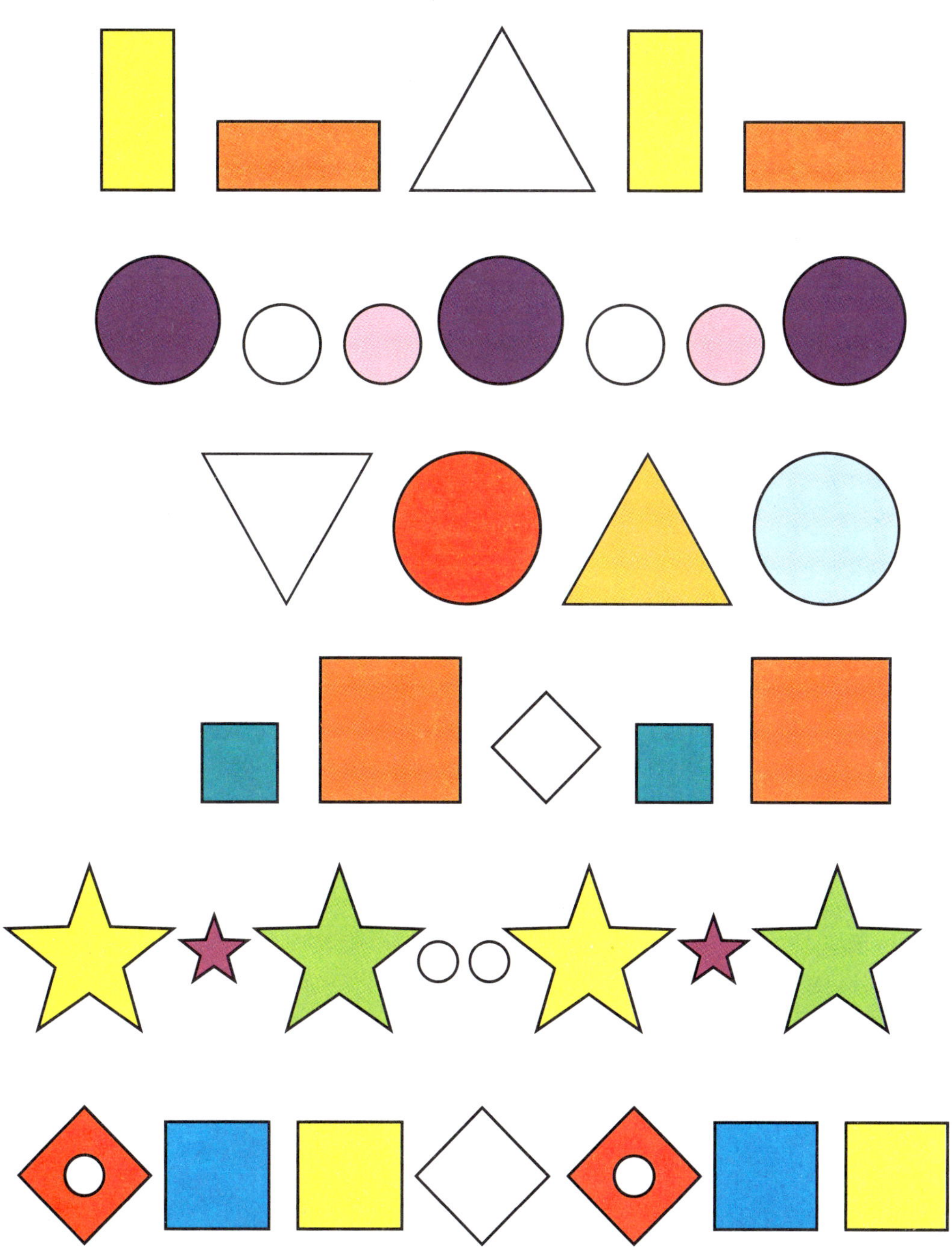

Patterns

Directions: Fill in the missing shape in each row. Then color it.

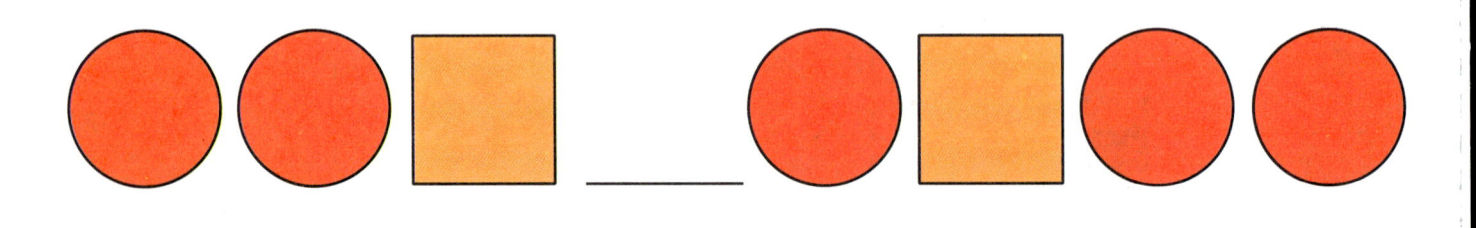

Summer Link Math Grade 1

Graphing

Directions: How many fire engines did the children have? Count the boxes. Write the number. How many of each vehicle? Count the boxes. Write the numbers.

Name _____

Graphing

Directions: Count the spots on the turtles. Color the boxes to show how many spots each turtle has.

1	2	3	4	5	6	7	8

1	2	3	4	5	6	7	8

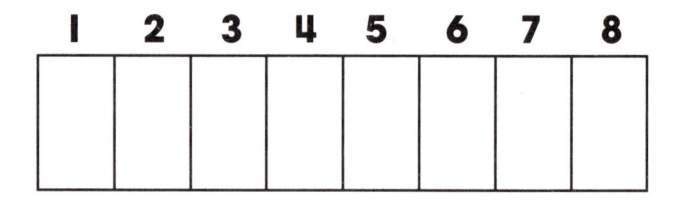

1	2	3	4	5	6	7	8

1	2	3	4	5	6	7	8

1	2	3	4	5	6	7	8

Name _____

Graphing

Directions: Count the shapes in the picture. Then complete the graph below.

Graphing

Directions: Look at the graph below. Then answer the questions.

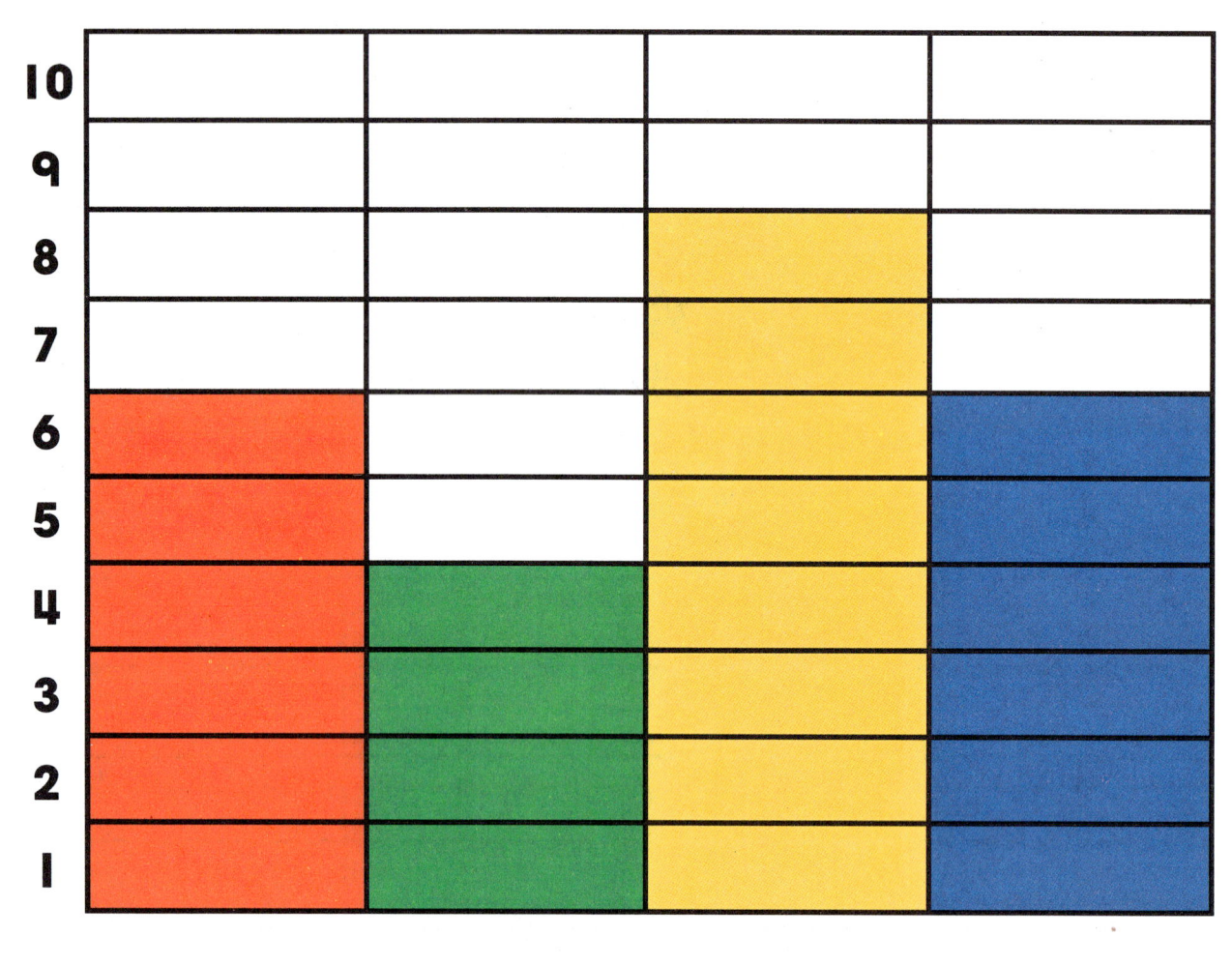

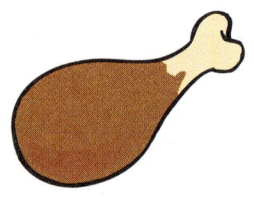

	hot dog	hamburger	pizza	chicken
10				
9				
8			■	
7			■	
6	■		■	■
5	■		■	■
4	■	■	■	■
3	■	■	■	■
2	■	■	■	■
1	■	■	■	■

Name _____

Graphing

◆ How many people like hot dogs best? _____

◆ How many people like pizza best? _____

◆ How many people like chicken best? _____

◆ Which food do most people like best? _____

◆ Which two foods do the same number of people like best?

_____ and _____

◆ Which food do the fewest number of people like best?

_____

Graphing

Directions: Count the pets in the window. Then color one box for each animal on the graph below.

Page 4
Number Recognition
Directions: Write the numbers 1-10. Color the bear.

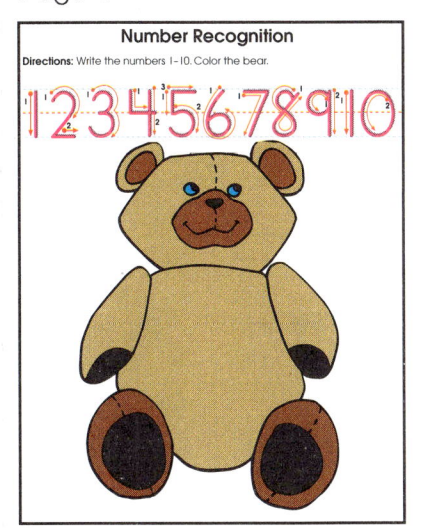

1 2 3 4 5 6 7 8 9 10

Page 5
Zero
Directions: Color the tank to show that it has 0 fish. Color the tanks that have 0 fish.

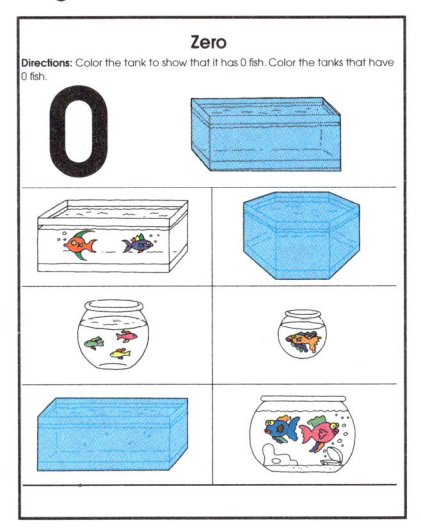

0

Page 6
One and Two
Directions: Count how many cars are on each track. Circle the number that shows how many.

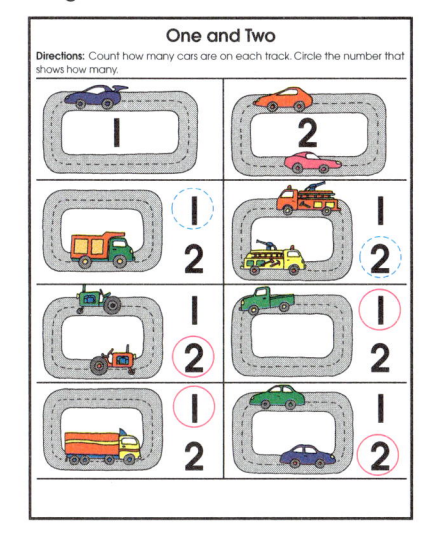

Page 7
Three
Directions: Color the 3 kittens in the basket. Color 3 animals in each group.

3

Page 8
Four
Directions: Color the 4 crayons. Count how many. Circle the correct number.

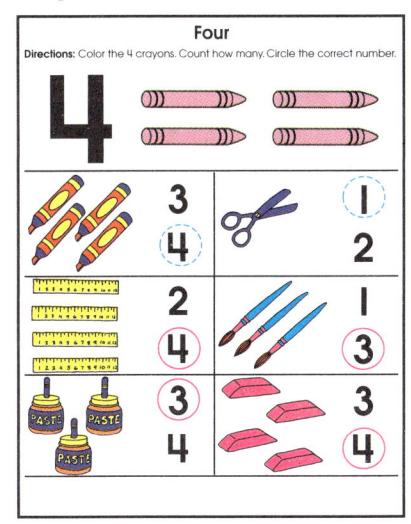

4

Page 9
Five
Directions: Color the 5 party hats. Color and circle the groups that have 5.

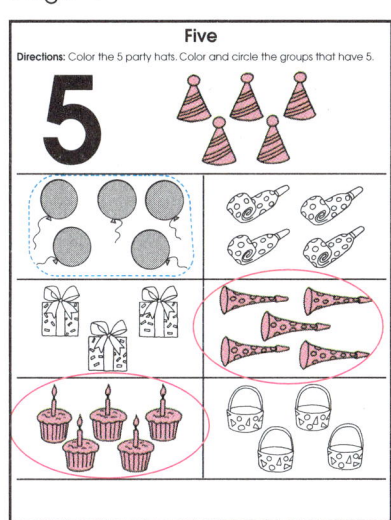

5

Page 10
Review Numbers 1-5
Directions: Look at the picture. Read the questions. Circle the correct number.

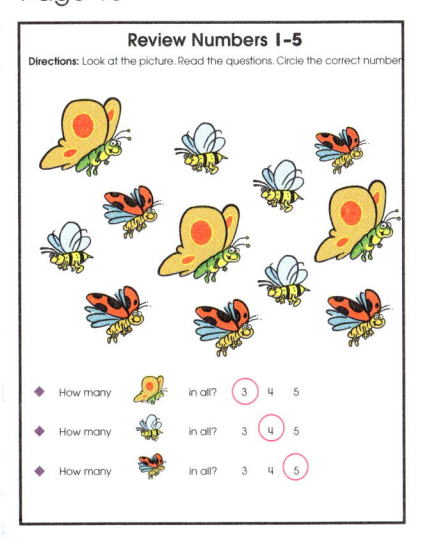

- How many ⚡ in all? 3 4 5
- How many 🐝 in all? 3 4 5
- How many 🦋 in all? 3 4 5

Page 11
Review Numbers 1-5
Directions: Draw a line from the number to the group that matches it.

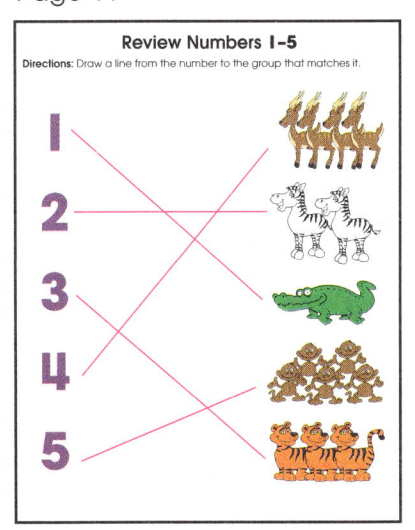

1
2
3
4
5

Page 12
Six
Directions: Look at the number 6. Count the teddy bears. Trace the circle to show this is a group of 6. Circle the group if it shows 6.

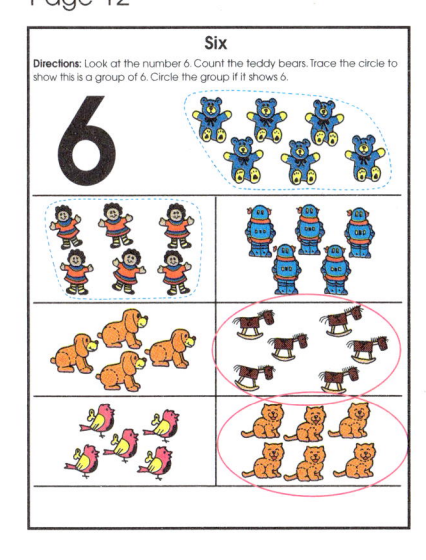

6

85

Summer Link Math Grade 1

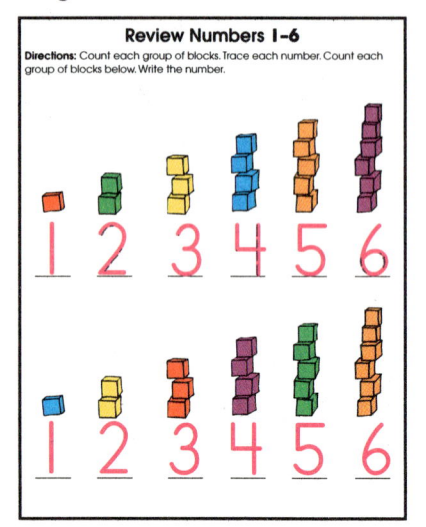

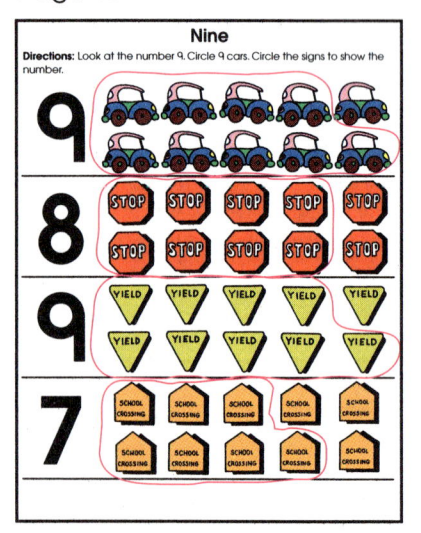

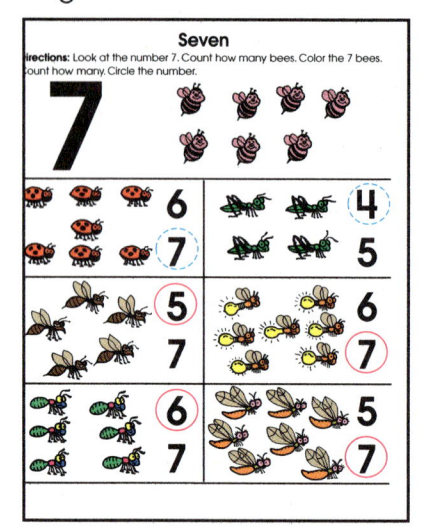

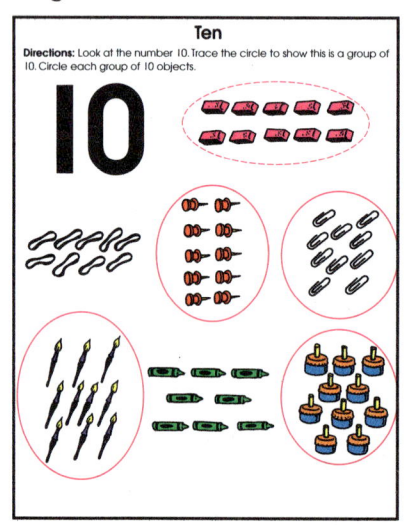

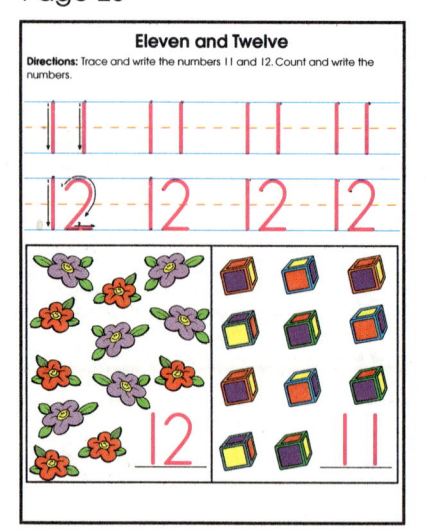

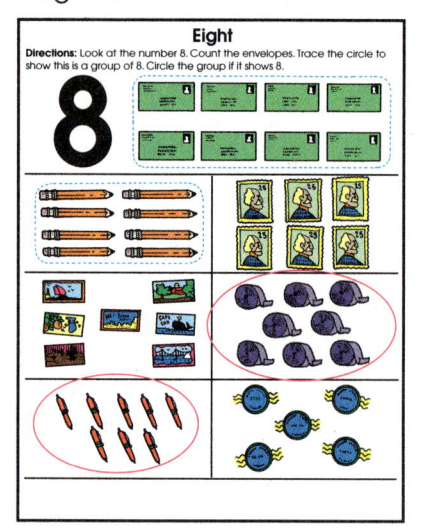

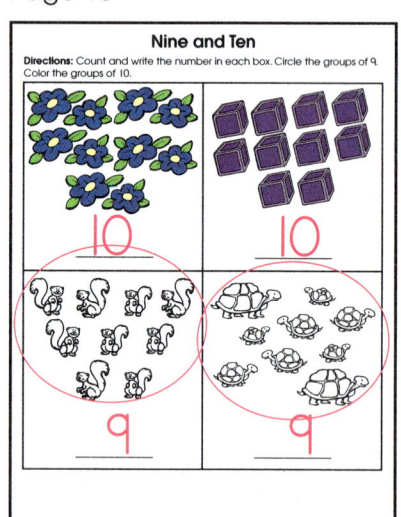

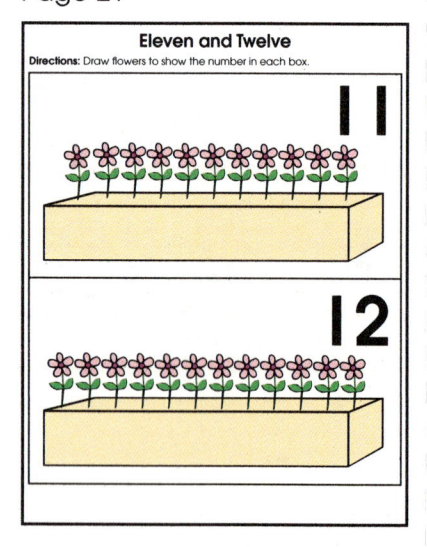

Page 13 · Page 14 · Page 15
Page 16 · Page 17 · Page 18
Page 19 · Page 20 · Page 21

Summer Link Math Grade 1

Page 22

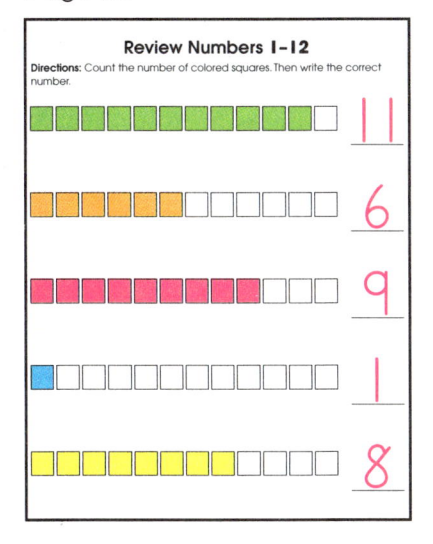

Review Numbers 1-12
Directions: Count the number of colored squares. Then write the correct number.

11
6
9
1
8

Page 23

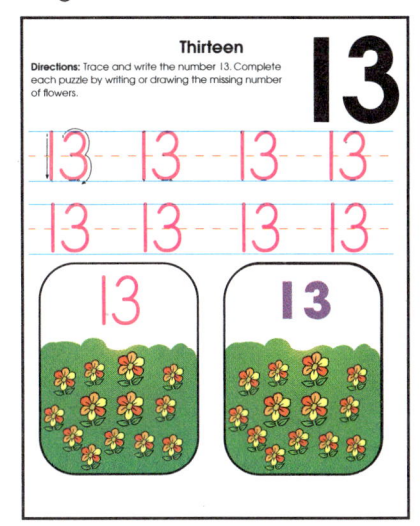

Thirteen 13
Directions: Trace and write the number 13. Complete each puzzle by writing or drawing the missing number of flowers.

13 13 13 13
13 13 13 13

Page 24

Fourteen 14
Directions: Trace and write the number 14. Connect the dots. Color the picture. What is it?

14 14 14 14
14 14 14 14

Page 25

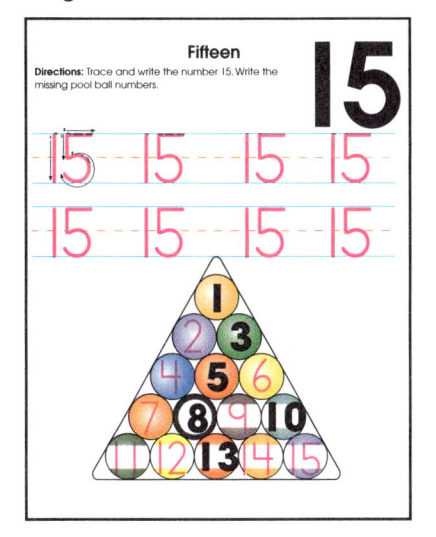

Fifteen 15
Directions: Trace and write the number 15. Write the missing pool ball numbers.

15 15 15 15
15 15 15 15

Page 26

Sixteen 16
Directions: Trace and write the number 16. Draw eight legs on each spider.

16 16 16 16
16 16 16 16

How many legs are there in all? _____

Page 27

Seventeen 17
Directions: Trace and write the number 17. Circle each group of 17 things. Color the dog.

17 17 17 17
17 17 17 17

Page 28

Eighteen 18
Directions: Trace and write the number 18. Help Filbert Fish find his way to the top. Write the numbers 1–18 in each bubble along the way.

18 18 18 18
18 18 18 18

Page 29

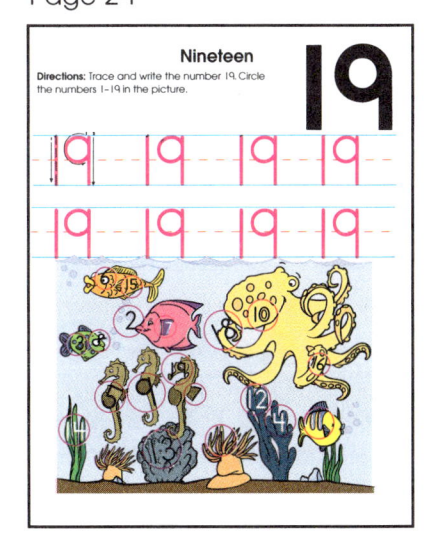

Nineteen 19
Directions: Trace and write the number 19. Circle the numbers 1–19 in the picture.

19 19 19 19
19 19 19 19

Page 30

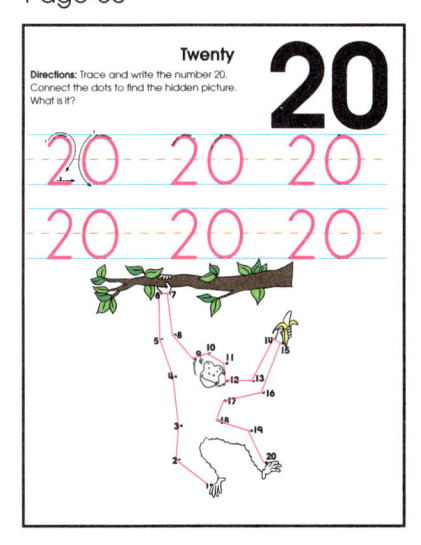

Twenty 20
Directions: Trace and write the number 20. Connect the dots to find the hidden picture. What is it?

20 20 20
20 20 20

87

Summer Link Math Grade 1

Summer Link Math Grade 1

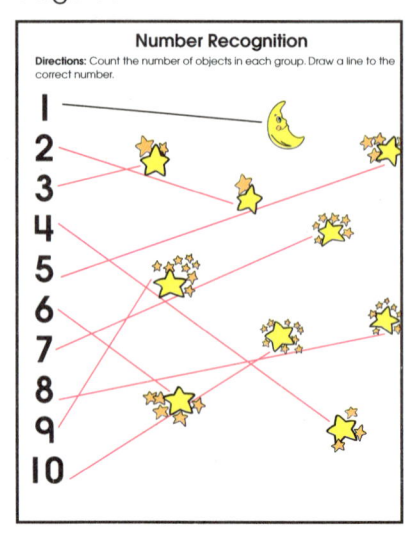

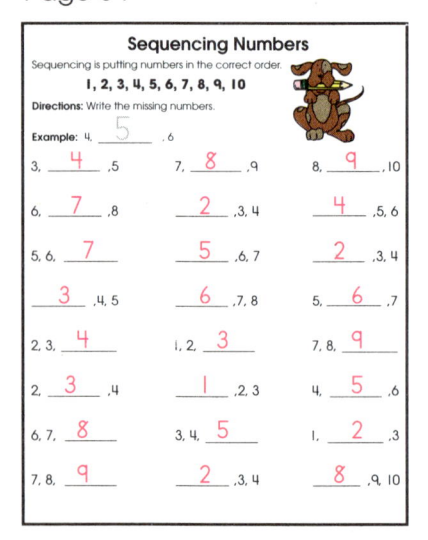

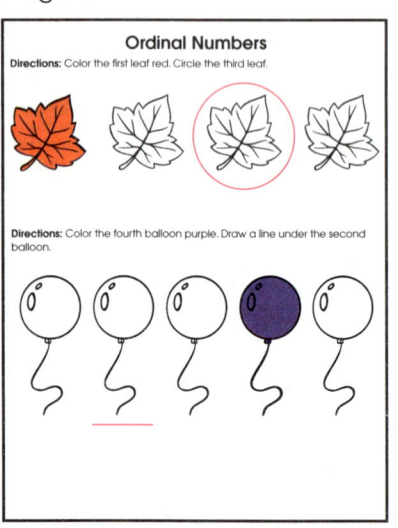

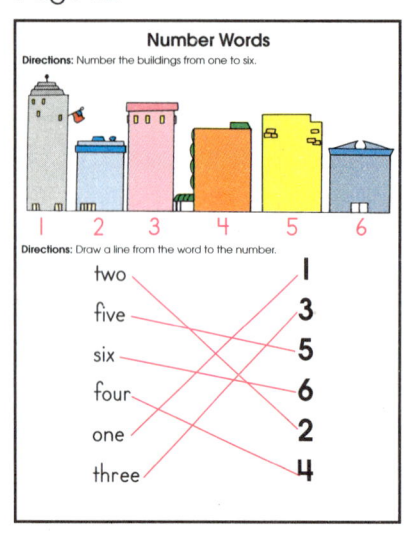

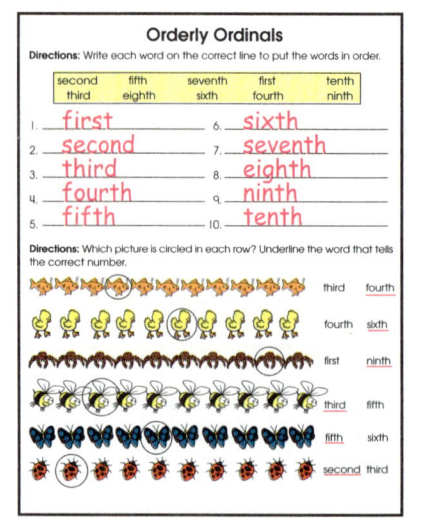

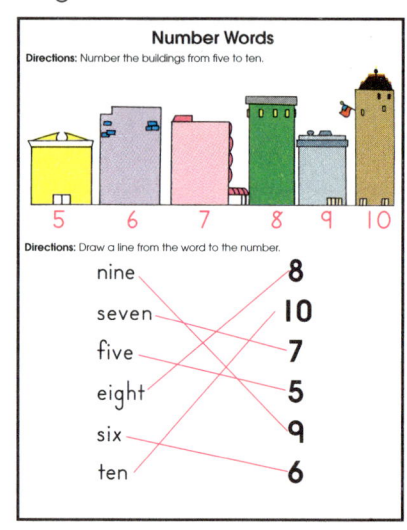

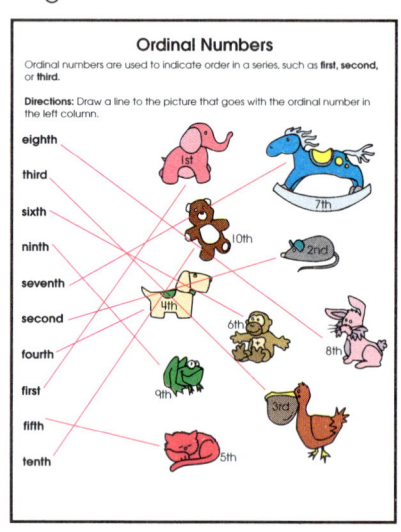

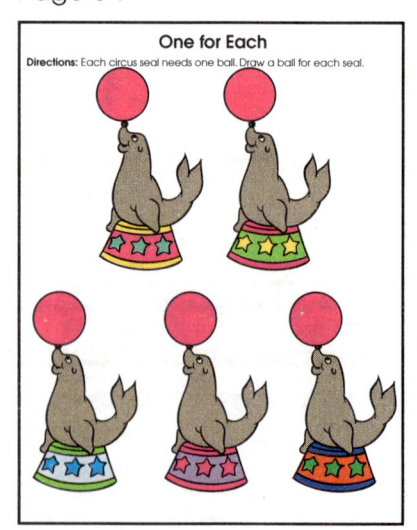

88

Page 40
More

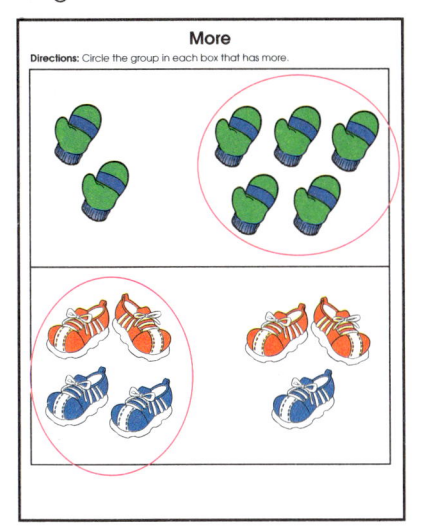

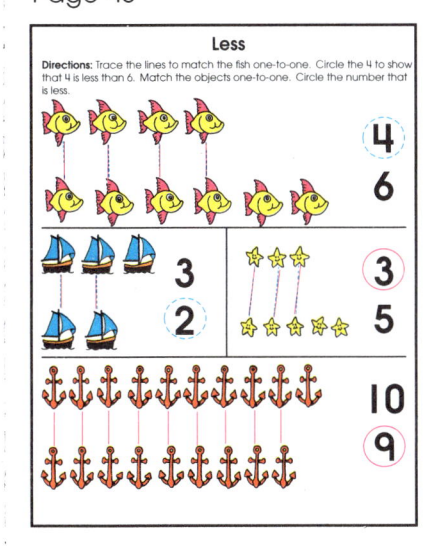

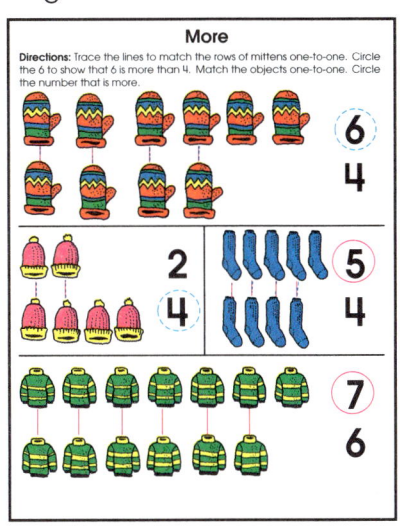

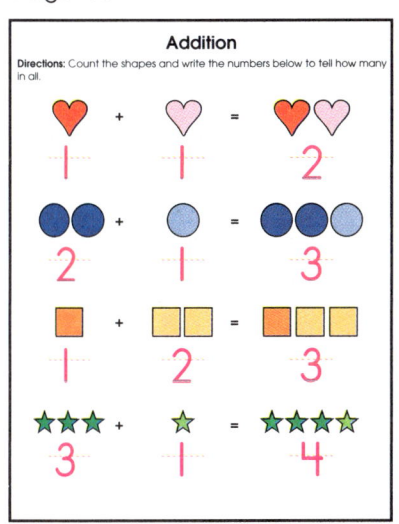

Page 41
More

Page 42
Less

Page 43
Less

Page 44
Addition 1, 2

Page 45
Addition

Page 46
Picture Problems: Addition

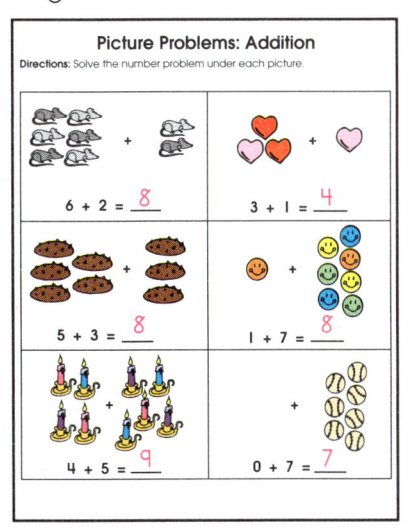

6 + 2 = 8
3 + 1 = 4
5 + 3 = 8
1 + 7 = 8
4 + 5 = 9
0 + 7 = 7

Page 47
How Many in All?

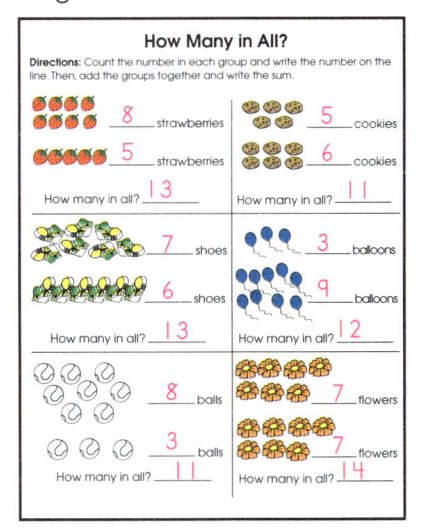

8 strawberries, 5 cookies
5 strawberries, 6 cookies
How many in all? 13   How many in all? 11

7 shoes, 3 balloons
6 shoes, 9 balloons
How many in all? 13   How many in all? 12

8 balls, 7 flowers
3 balls, 7 flowers
How many in all? 11   How many in all? 14

Page 48
Subtraction 1, 2, 3

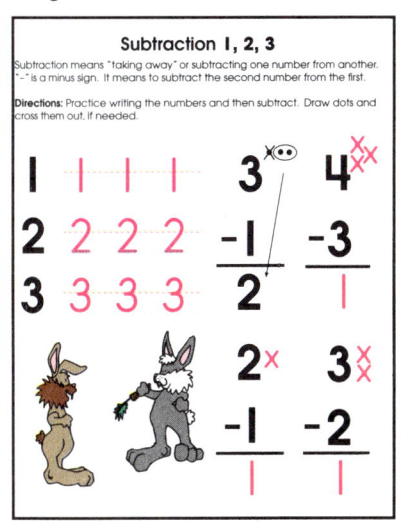

89

Summer Link Math Grade 1

Page 49

Picture Problems: Subtraction

Directions: Solve the number problem under each picture.

5 − 2 = 3   6 − 1 = 5
7 − 4 = 3   8 − 3 = 5
9 − 2 = 7   4 − 4 = 0

Page 50

Picture Problems: Addition and Subtraction

Directions: Solve the number problem under each picture.

7 − 4 = 3   1 + 4 = 5
3 + 5 = 8   8 − 1 = 7
9 + 5 = 14   6 − 3 = 3

Page 51

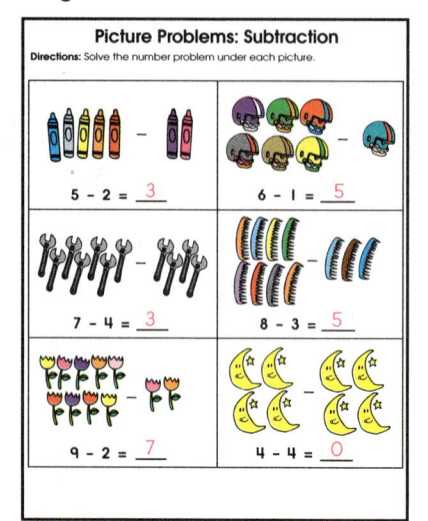

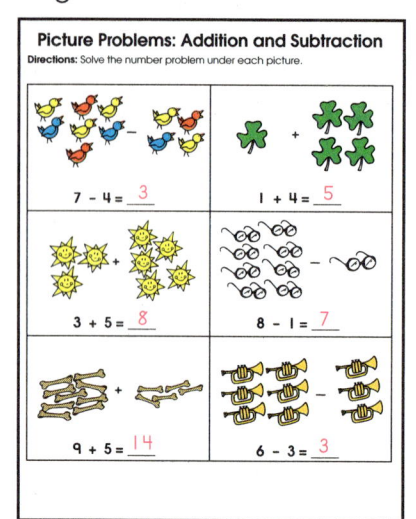

Review: Addition and Subtraction

Directions: Solve the number problem under each picture. Write + or − to show if you should add or subtract.

How many are left? How many in all?
12 − 4 = 8   6 + 8 = 14

How many are left? How many are left?
4 − 4 = 0   11 − 7 = 4

How many in all?   How many in all?
9 + 3 = 12   10 + 0 = 10

Page 52

Addition and Subtraction

Directions: Solve the problems. Remember, addition means "putting together" or adding two or more numbers to find the sum. Subtraction means "taking away" or subtracting one number from another.

1 + 3 = 4   4 − 3 = 1   4 + 5 = 9
6 + 1 = 7   7 − 2 = 5   8 − 4 = 4
9 − 1 = 8   10 − 3 = 7
5 − 2 = 3   6 + 3 = 9
8 + 2 = 10   5 + 5 = 10

Page 53

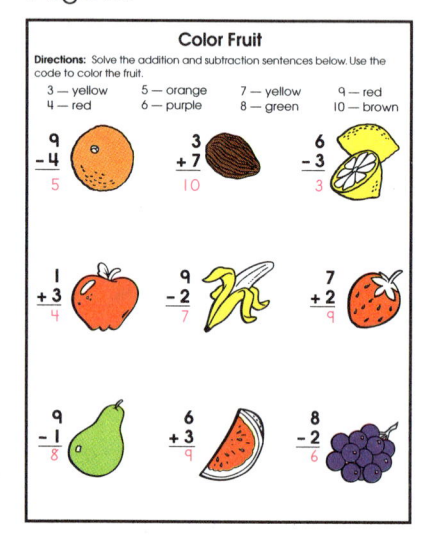

Color Fruit

Directions: Solve the addition and subtraction sentences below. Use the code to color the fruit.

3 — yellow   5 — orange   7 — yellow   9 — red
4 — red   6 — purple   8 — green   10 — brown

9 − 4 = 5   3 + 7 = 10   6 − 3 = 3
1 + 3 = 4   9 − 2 = 7   7 + 2 = 9
9 − 1 = 8   6 + 3 = 9   8 − 2 = 6

Page 54

Time

Directions: Trace the numbers 1–12 in order on the clock.

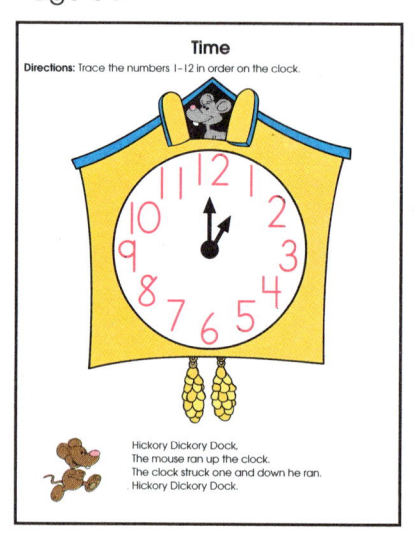

Hickory Dickory Dock,
The mouse ran up the clock.
The clock struck one and down he ran.
Hickory Dickory Dock.

Page 55

Time

Directions: Write the time that is on each clock.

Example: 2 o'clock
3 o'clock
9 o'clock
6 o'clock

Page 56

Time

Directions: Write the time that is on each clock.

10 o'clock
11 o'clock
7 o'clock
8 o'clock

Page 57

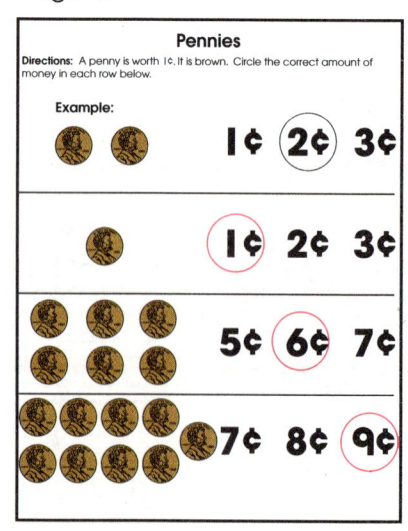

Pennies

Directions: A penny is worth 1¢. It is brown. Circle the correct amount of money in each row below.

Example: 1¢ (2¢) 3¢
1¢ 2¢ 3¢
5¢ (6¢) 7¢
7¢ 8¢ (9¢)

Summer Link Math Grade 1

Page 58

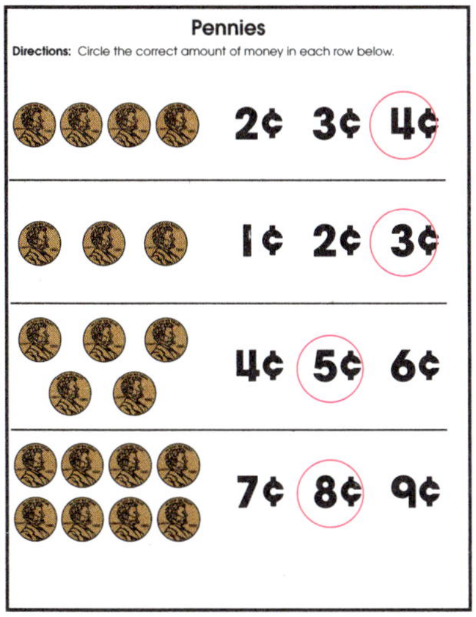

Pennies
Directions: Circle the correct amount of money in each row below.

- 2¢   3¢   **4¢**
- 1¢   2¢   **3¢**
- 4¢   **5¢**   6¢
- 7¢   **8¢**   9¢

Page 59

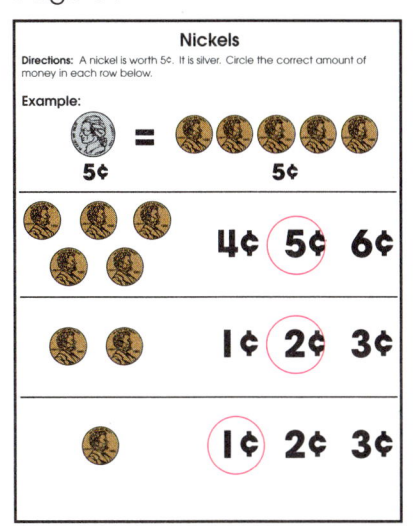

Nickels
Directions: A nickel is worth 5¢. It is silver. Circle the correct amount of money in each row below.

Example: 5¢ = 5¢

- 4¢   **5¢**   6¢
- 1¢   **2¢**   3¢
- **1¢**   2¢   3¢

Page 60

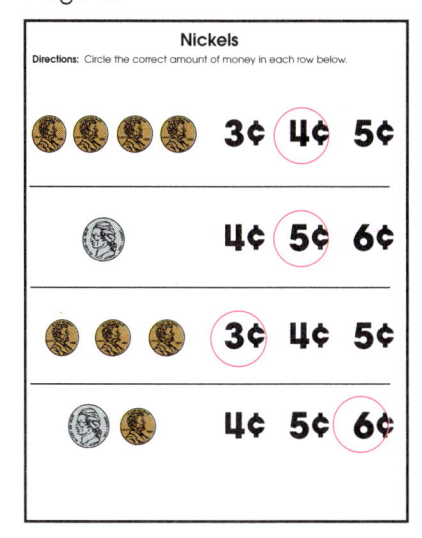

Nickels
Directions: Circle the correct amount of money in each row below.

- 3¢   **4¢**   5¢
- 4¢   **5¢**   6¢
- **3¢**   4¢   5¢
- 4¢   5¢   **6¢**

Page 61

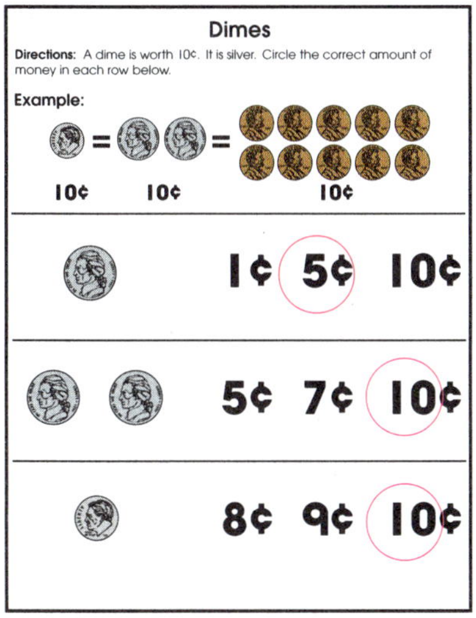

Dimes
Directions: A dime is worth 10¢. It is silver. Circle the correct amount of money in each row below.

Example: 10¢ = 10¢ = 10¢

- 1¢   **5¢**   10¢
- 5¢   7¢   **10¢**
- 8¢   9¢   **10¢**

Page 62

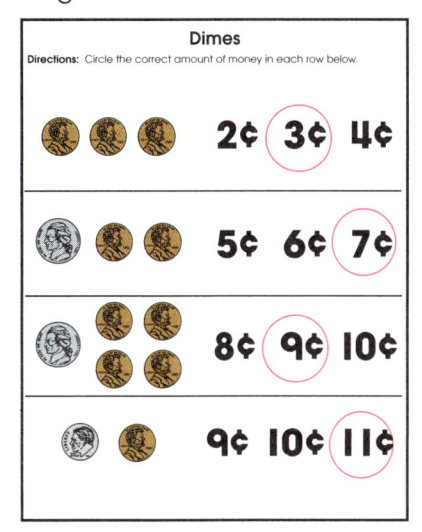

Dimes
Directions: Circle the correct amount of money in each row below.

- 2¢   **3¢**   4¢
- 5¢   6¢   **7¢**
- 8¢   **9¢**   10¢
- 9¢   **10¢**   11¢

Page 63

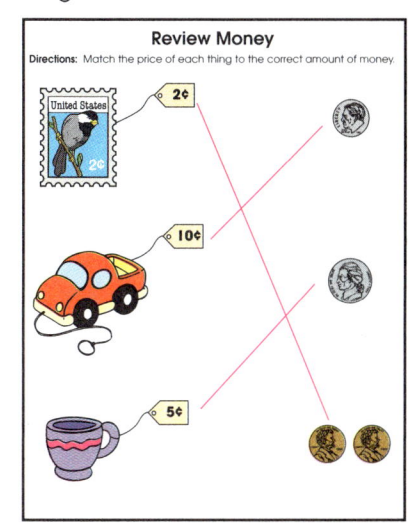

Review Money
Directions: Match the price of each thing to the correct amount of money.

Page 64

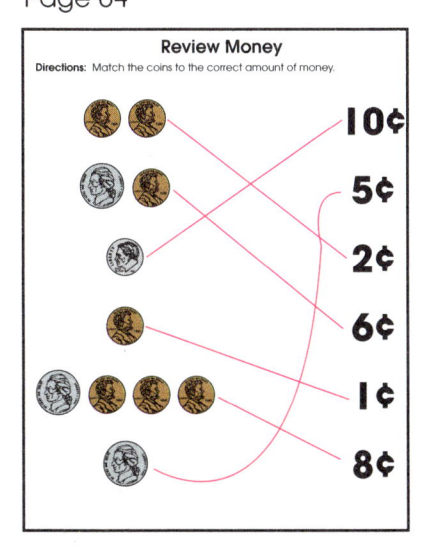

Review Money
Directions: Match the coins to the correct amount of money.

- 10¢
- 5¢
- 2¢
- 6¢
- 1¢
- 8¢

Page 65

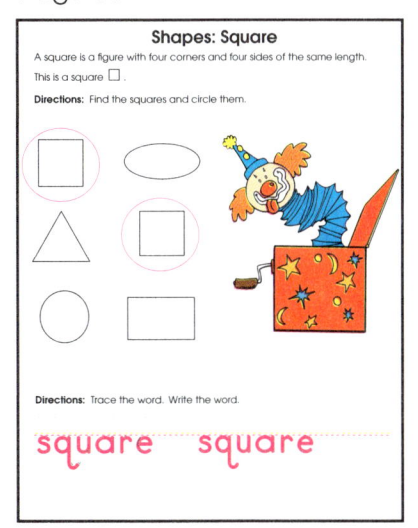

Shapes: Square
A square is a figure with four corners and four sides of the same length. This is a square ☐.

Directions: Find the squares and circle them.

Directions: Trace the word. Write the word.

square   square

Page 66

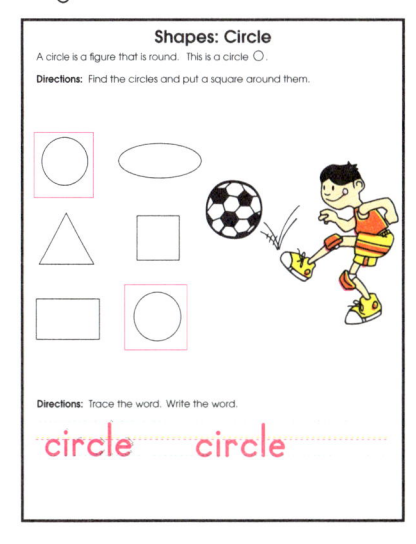

Shapes: Circle
A circle is a figure that is round. This is a circle ○.

Directions: Find the circles and put a square around them.

Directions: Trace the word. Write the word.

circle   circle

Page 67

Shapes: Square and Circle

Directions: Practice drawing squares. Trace the samples and make four of your own.

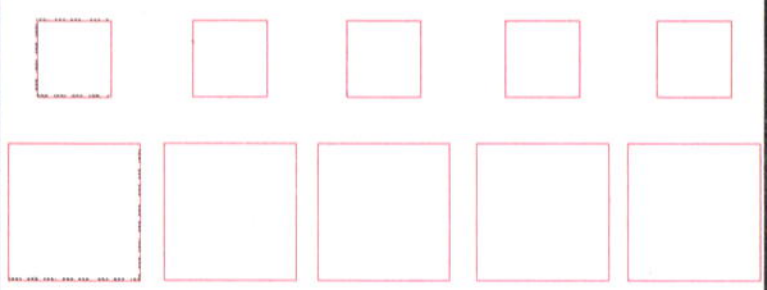

Directions: Practice drawing circles. Trace the samples and make four of your own.

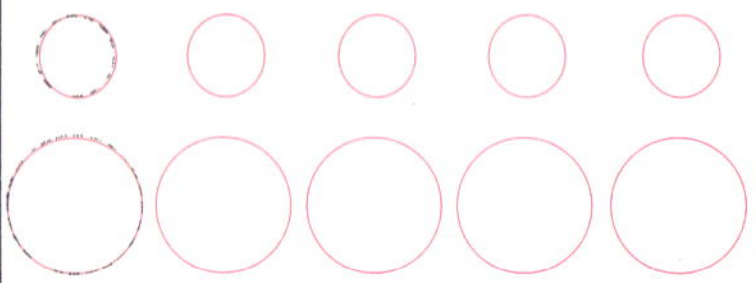

Page 68

Shapes: Triangle

A triangle is a figure with three corners and three sides.

This is a triangle △.

Directions: Find the triangles and put a circle around them.

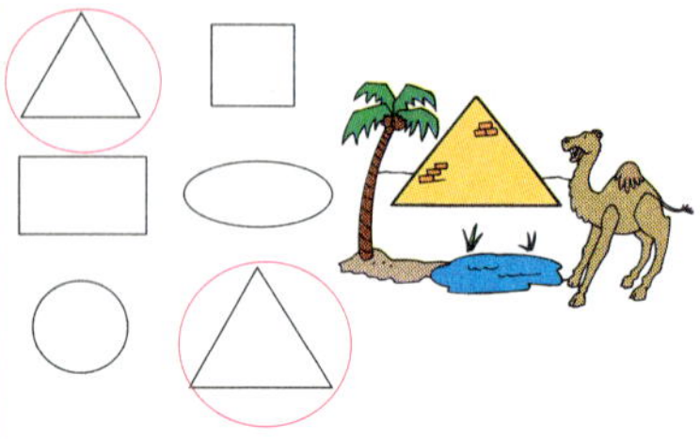

Directions: Trace the word. Write the word.

triangle triangle

Page 69

Shapes: Rectangle

A rectangle is a figure with four corners and four sides. Sides opposite each other are the same length.

This is a rectangle ▭.

Directions: Find the rectangles and put a circle around them.

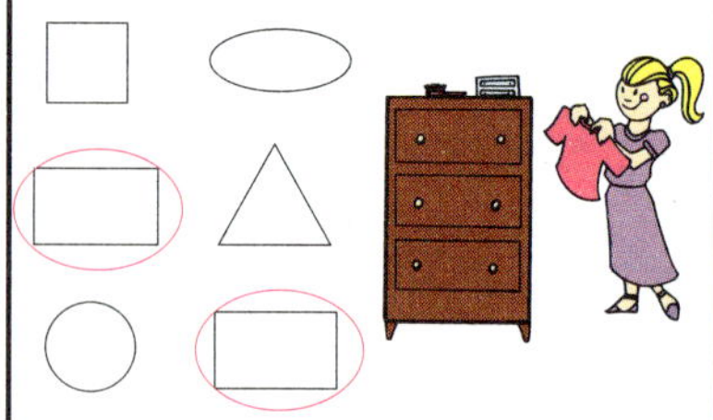

Directions: Trace the word. Write the word.

rectangle rectangle

Page 70

Shapes: Triangle and Rectangle

Directions: Practice drawing triangles. Trace the samples and make four of your own.

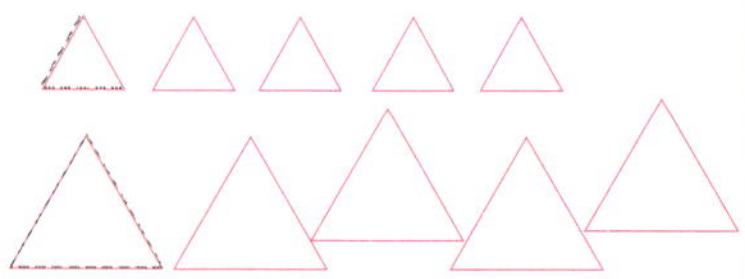

Directions: Practice drawing rectangles. Trace the samples and make four of your own.

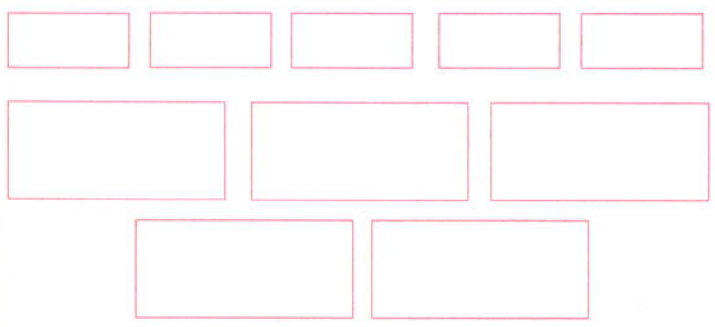

Page 71

Shapes: Oval and Diamond

An oval is an egg-shaped figure. A diamond is a figure with four sides of the same length. Its corners form points at the top, sides, and bottom.

This is an oval ⬭. This is a diamond ◇.

Directions: Color the ovals red. Color the diamonds blue.

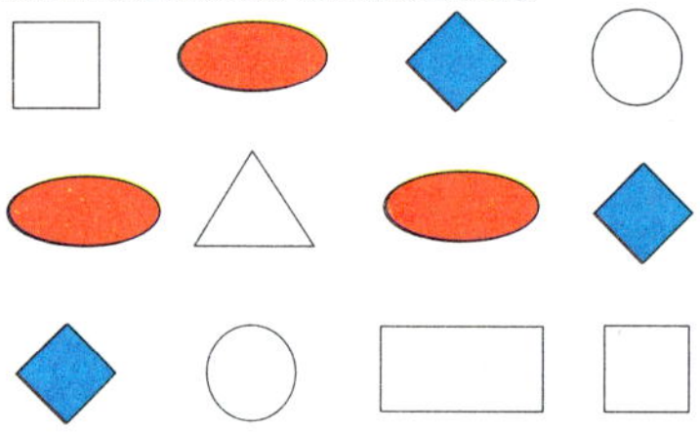

Directions: Trace the words. Write the words.

oval oval

diamond diamond

Page 72

Shapes: Oval and Diamond

Directions: Practice drawing ovals. Trace the samples and make four of your own.

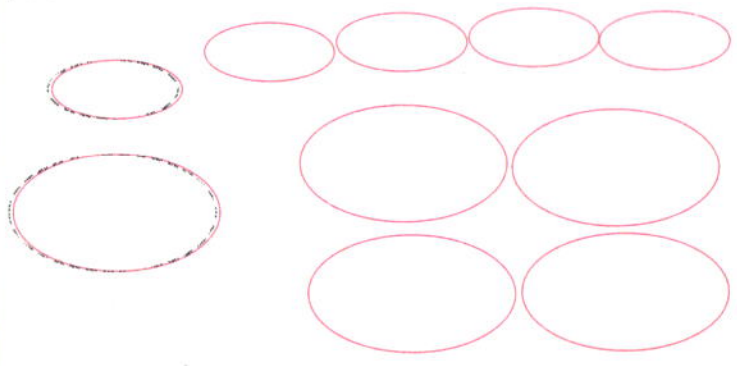

Directions: Practice drawing diamonds. Trace the samples and make four of your own.

Page 73

Animal Shapes

Directions: Color: squares — green rectangles — yellow circles — red triangles — blue

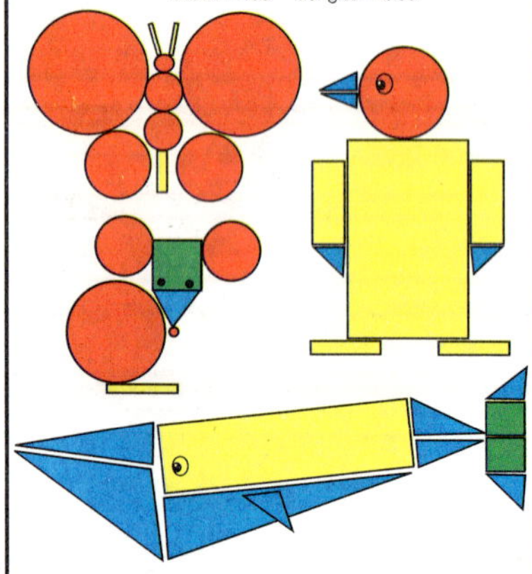

Page 74

Classifying: Shapes

Mary and Rudy are taking a trip into space. Help them find the stars, moons, circles, and diamonds.

Directions: Color the shapes.
Use yellow for ☆s. Use blue for ☾s.
Use red for ○s. Use purple for ◇s.

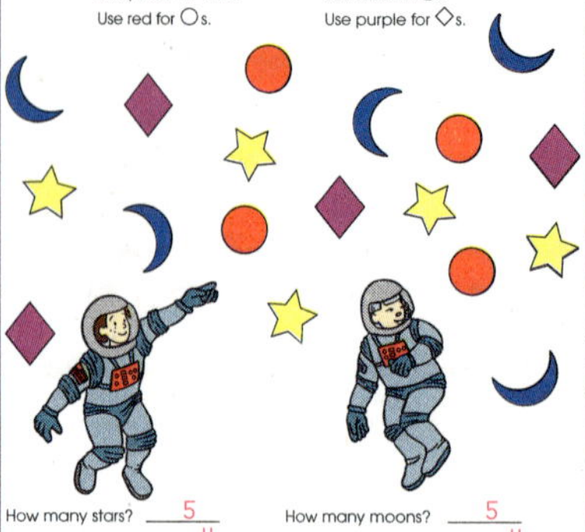

How many stars? 5
How many circles? 4
How many moons? 5
How many diamonds? 4

Page 75

Same Shape

Directions: Look at the round shapes. They are all the same shape. Draw a line from each shape in the bottom row to the box with the same shape.

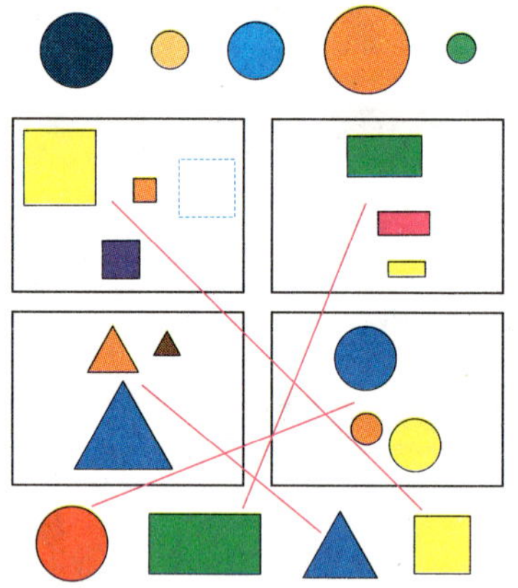

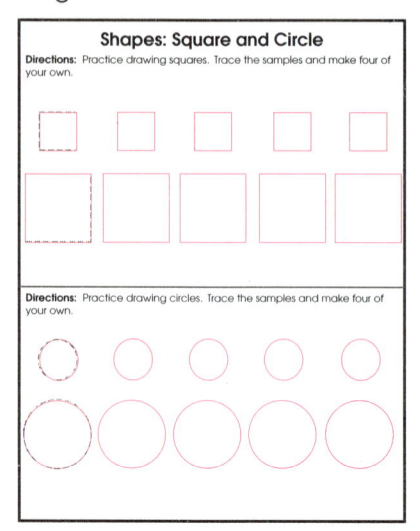

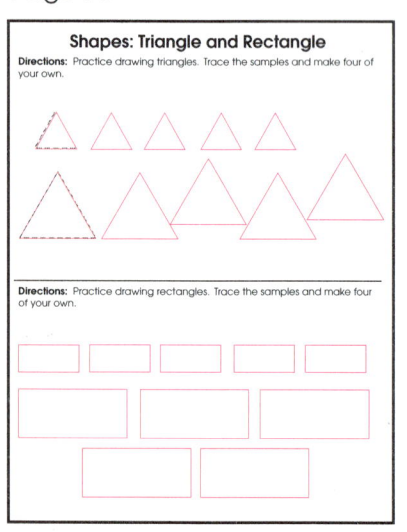

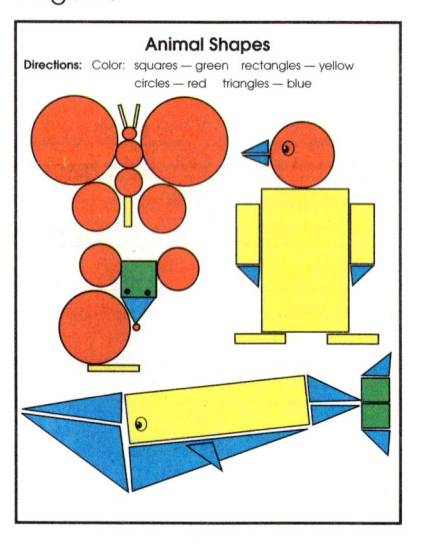

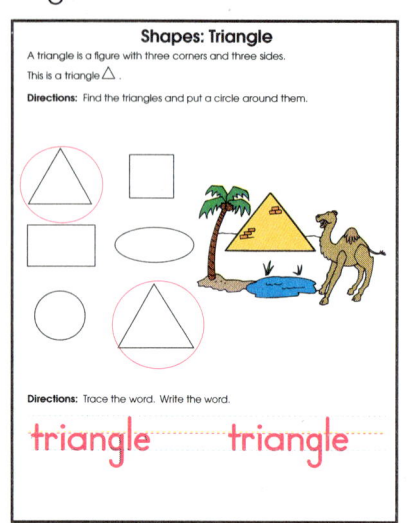

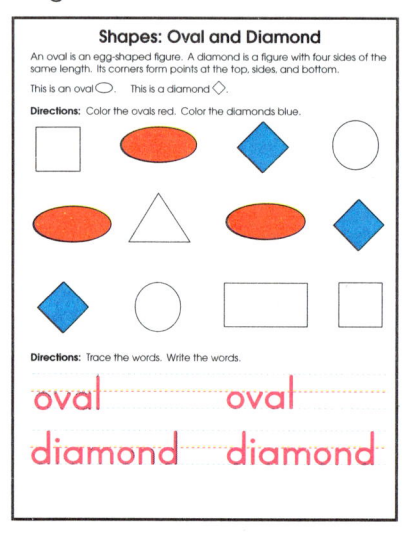

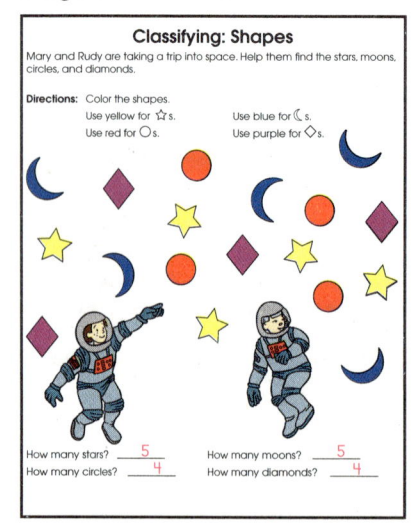

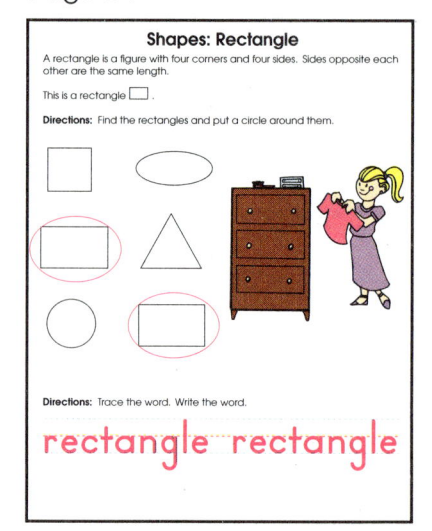

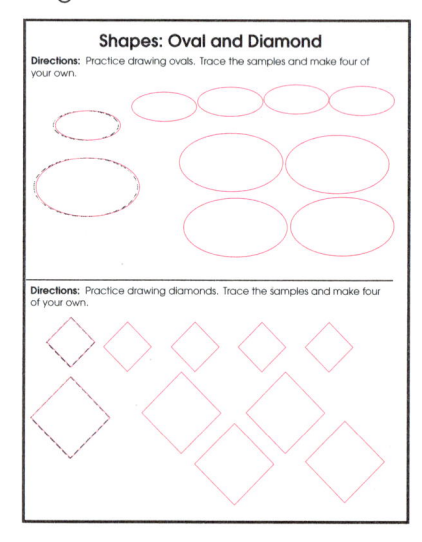

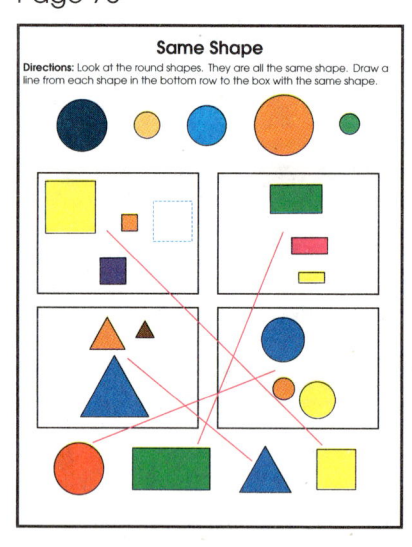

Page 76

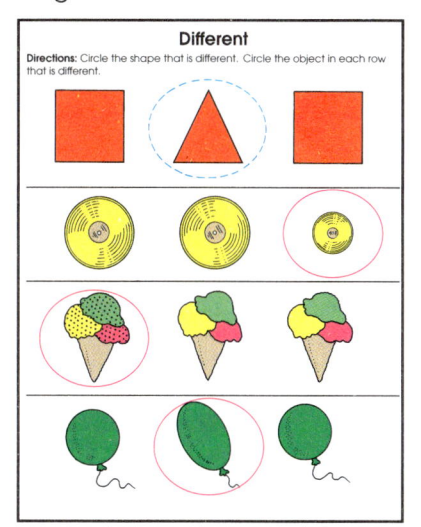

Page 77

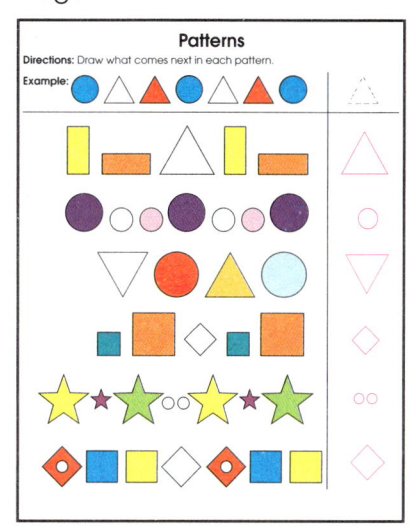

Page 78

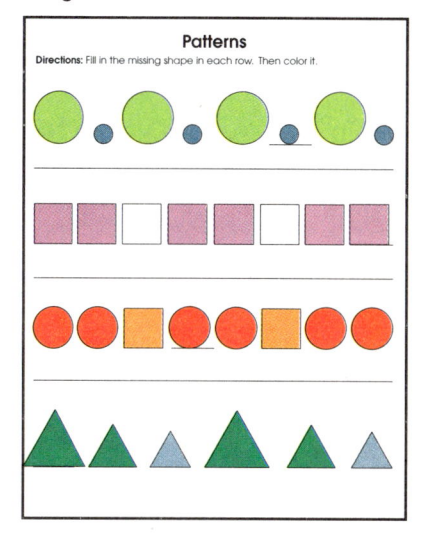

Page 79

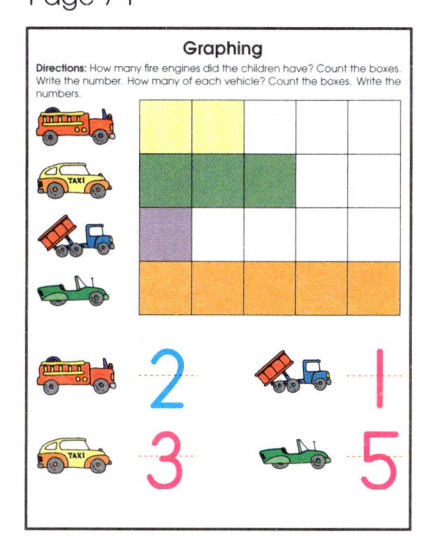

Page 80

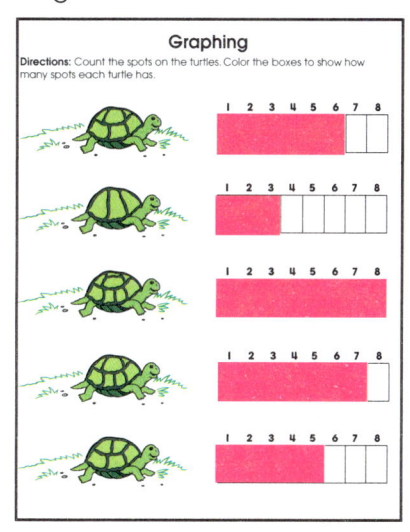

Page 81

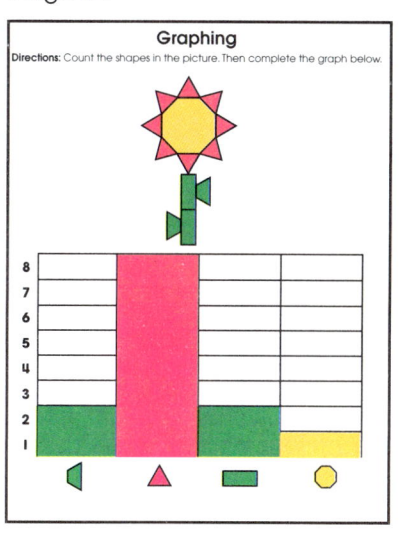

Page 83
- How many people like hot dogs best? **6**
- How many people like pizza best? **8**
- How many people like chicken best? **6**
- Which food do most people like best? **pizza**
- Which two foods do the same number of people like best? **hot dogs** and **chicken**
- Which food do the fewest number of people like best? **hamburgers**

Page 84

93

Summer Link Math Grade 1

Developmental Skills for First Grade Success

This checklist is designed to help you assess your child's progress in the following kindergarten skills. You may want to add to or adapt this checklist to fit your child's abilities.

Basic Skills

- [ ] Names basic colors _____
- [ ] Names simple shapes _____
- [ ] Identifies opposites _____
- [ ] Understands positional concepts _____
- [ ] Names days of the week in order _____

Mathematics Readiness

- [ ] Counts objects to 20 _____
- [ ] Writes numbers to 20 _____
- [ ] Identifies numbers to 20 in random order _____
- [ ] Rote counts to 100 _____
- [ ] Counts by 10's to 100 _____
- [ ] Uses ordinal numbers _____
- [ ] Reads a graph _____
- [ ] Identifies and continues established patterns _____

Fine (Small) Motor Skills

- [ ] Colors within lines _____
- [ ] Draws shapes _____
- [ ] Holds a pencil _____
- [ ] Prints letters and numbers _____
- [ ] Cuts a line with scissors _____

SPECTRUM

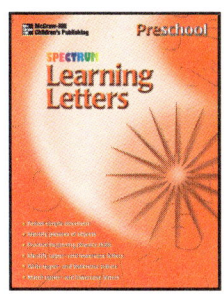

PRESCHOOL

Learning Letters offers comprehensive instruction and practice in following directions, recognizing and writing upper- and lowercase letters, and beginning phonics. Math Readiness features activities that teach such important skills as counting, identifying numbers, creating patterns, and recognizing "same and different." Basic Concepts and Skills offers exercises that help preschoolers identify colors, read and write words, identify simple shapes, and more. 160 pages.

TITLE	ISBN	PRICE
Learning Letters	1-57768-329-3	$8.95
Math Readiness	1-57768-339-0	$8.95
Basic Concepts and Skills	1-57768-349-8	$8.95

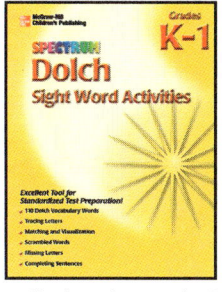

DOLCH SIGHT WORD ACTIVITIES

The Dolch Sight Word Activities workbooks use the classic Dolch list of 220 basic vocabulary words that make up from 50 to 75 percent of all reading matter that children ordinarily encounter. Since these words are ordinarily recognized on sight, they are called *sight words*. Volume 1 includes 110 sight words. Volume 2 covers the remainder of the list. 160 pages. Answer key included.

TITLE	ISBN	PRICE
Grades K-1 Vol. 1	1-56189-917-8	$9.95
Grades K-1 Vol. 2	1-56189-918-6	$9.95

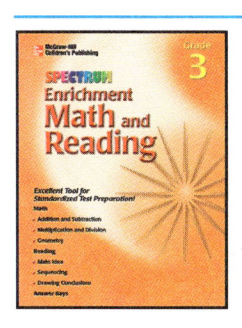

ENRICHMENT MATH AND READING

Books in this series offer advanced math and reading for students excelling in grades 3–6. Lessons follow the same curriculum children are being taught in school while presenting the material in a way that children feel challenged. 160 pages. Answer key included.

TITLE	ISBN	PRICE
Grade 3	1-57768-503-2	$8.95
Grade 4	1-57768-504-0	$8.95
Grade 5	1-57768-505-9	$8.95
Grade 6	1-57768-506-7	$8.95

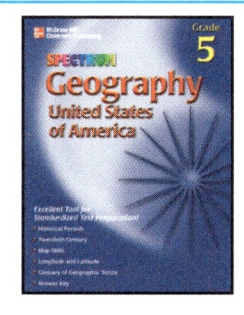

GEOGRAPHY

Full-color, three-part lessons strengthen geography knowledge and map-reading skills. Focusing on five geographic themes including location, place, human/environmental interaction, movement, and regions. Over 150 pages. Glossary of geographical terms and answer key included.

TITLE	ISBN	PRICE
Grade 3, Communities	1-56189-963-1	$8.95
Grade 4, Regions	1-56189-964-X	$8.95
Grade 5, USA	1-56189-965-8	$8.95
Grade 6, World	1-56189-966-6	$8.95

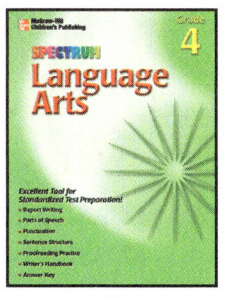

LANGUAGE ARTS

Encourages creativity and builds confidence by making writing fun! Seventy-two four-part lessons strengthen writing skills by focusing on parts of speech, word usage, sentence structure, punctuation, and proofreading. Each book includes a Writer's Handbook at the end of the book that offers writing tips. This series is based on the highly respected SRA/McGraw-Hill language arts series. More than 180 full-color pages. Answer key included.

TITLE	ISBN	PRICE
Grade 2	1-56189-952-6	$8.95
Grade 3	1-56189-953-4	$8.95
Grade 4	1-56189-954-2	$8.95
Grade 5	1-56189-955-0	$8.95
Grade 6	1-56189-956-9	$8.95

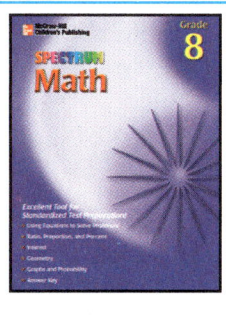

MATH

Features easy-to-follow instructions that give students a clear path to success. This series has comprehensive coverage of the basic skills, helping children to master math fundamentals. Over 150 pages. Answer key included.

TITLE	ISBN	PRICE
Grade K	1-56189-900-3	$8.95
Grade 1	1-56189-901-1	$8.95
Grade 2	1-56189-902-X	$8.95
Grade 3	1-56189-903-8	$8.95
Grade 4	1-56189-904-6	$8.95
Grade 5	1-56189-905-4	$8.95
Grade 6	1-56189-906-2	$8.95
Grade 7	1-56189-907-0	$8.95
Grade 8	1-56189-908-9	$8.95

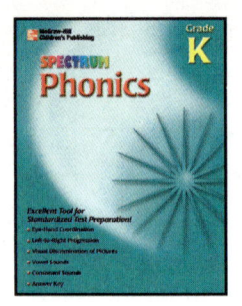

PHONICS/WORD STUDY

Provides everything children need to build multiple skills in language. Focusing on phonics, structural analysis, and dictionary skills, this series also offers creative ideas for using phonics and word study skills in other language areas. Over 200 pages. Answer key included.

TITLE	ISBN	PRICE
Grade K	1-56189-940-2	$8.95
Grade 1	1-56189-941-0	$8.95
Grade 2	1-56189-942-9	$8.95
Grade 3	1-56189-943-7	$8.95
Grade 4	1-56189-944-5	$8.95
Grade 5	1-56189-945-3	$8.95
Grade 6	1-56189-946-1	$8.95

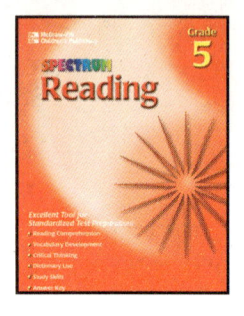

READING

This full-color series creates an enjoyable reading environment, even for below-average readers. Each book contains captivating content, colorful characters, and compelling illustrations, so children are eager to find out what happens next. Over 150 pages. Answer key included.

TITLE	ISBN	PRICE
Grade K	1-56189-910-0	$8.95
Grade 1	1-56189-911-9	$8.95
Grade 2	1-56189-912-7	$8.95
Grade 3	1-56189-913-5	$8.95
Grade 4	1-56189-914-3	$8.95
Grade 5	1-56189-915-1	$8.95
Grade 6	1-56189-916-X	$8.95

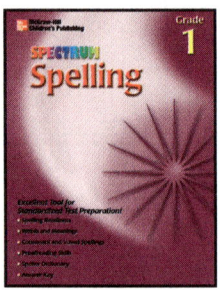

SPELLING

This full-color series links spelling to reading and writing, and increases skills in words and meanings, consonant and vowel spellings, and proofreading practice. Over 200 pages. Speller dictionary and answer key included.

TITLE	ISBN	PRICE
Grade 1	1-56189-921-6	$8.95
Grade 2	1-56189-922-4	$8.95
Grade 3	1-56189-923-2	$8.95
Grade 4	1-56189-924-0	$8.95
Grade 5	1-56189-925-9	$8.95
Grade 6	1-56189-926-7	$8.95

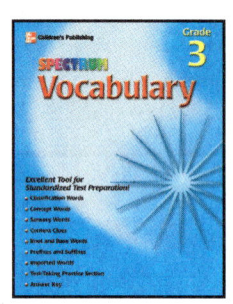

VOCABULARY

An essential building block for writing and reading proficiency, this series extends vocabulary knowledge through key concepts based on language arts and reading standards, offering a solid foundation for language arts, spelling, and reading comprehension. The series features a proficiency test practice section for standards-aligned assessment. Over 150 pages. Answer key included.

TITLE	ISBN	PRICE
Grade 3	1-57768-903-8	$8.95
Grade 4	1-57768-904-6	$8.95
Grade 5	1-57768-905-4	$8.95
Grade 6	1-57768-906-2	$8.95

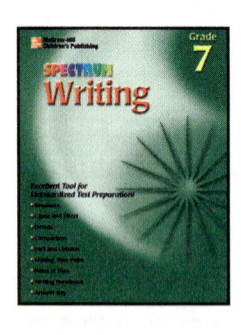

WRITING

Lessons focus on creative and expository writing using clearly stated objectives and pre-writing exercises. Eight essential reading skills are applied. Activities include main idea, sequence, comparison, detail, fact and opinion, cause and effect, making a point, and point of view. Over 130 pages. Answer key included.

TITLE	ISBN	PRICE
Grade 1	1-56189-931-3	$8.95
Grade 2	1-56189-932-1	$8.95
Grade 3	1-56189-933-X	$8.95
Grade 4	1-56189-934-8	$8.95
Grade 5	1-56189-935-6	$8.95
Grade 6	1-56189-936-4	$8.95
Grade 7	1-56189-937-2	$8.95
Grade 8	1-56189-938-0	$8.95

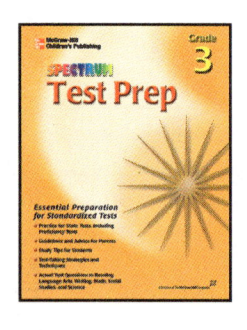

TEST PREP

Prepares children to do their best on current editions of the five major standardized tests. Activities reinforce test-taking skills through examples, tips, practice, and timed exercises. Subjects include reading, math, language arts, writing, social studies, and science. Over 150 pages. Answer key included.

TITLE	ISBN	PRICE
Grades 1-2	1-57768-672-1	$9.95
Grade 3	1-57768-673-X	$9.95
Grade 4	1-57768-674-8	$9.95
Grade 5	1-57768-675-6	$9.95
Grade 6	1-57768-676-4	$9.95
Grade 7	1-57768-677-2	$9.95
Grade 8	1-57768-678-0	$9.95

All our workbooks meet school curriculum guidelines and correspond to The McGraw-Hill Companies' classroom textbooks. Prices subject to change without notice.